• LOCOMOTIVES OF THE •

LANCASHIRE CENTRAL COALFIELD

THE WALKDEN YARD CONNECTION

ALAN DAVIES

AMBERLEY

First published 2014

Amberley Publishing
The Hill, Stroud
Gloucestershire, GL5 4EP

www.amberley-books.com

British Library Cataloguing in Publication Data.
A catalogue record for this book is available from the British Library.

ISBN 978 1 4456 3483 8
Ebook ISBN 978 1 4456 3504 0

Typeset in 10pt on 12pt Sabon.
Typesetting and Origination by Amberley Publishing.
Printed in the UK.

Contents

Acknowledgements

Thanks go to the late Joe Cunliffe (*d.* Sept 1993), former Superintendent at Walkden Yard, for allowing me to copy in the 1990s many of the photographs and loco maintenance documents formerly held there. Philip Hindley, railway historian and photographer, has been extremely generous and helpful in ensuring that facts, figures and technical information were accurate. Thanks to Glen Atkinson, historian of the Duke of Bridgewater's underground and surface canals and the Walkden area. Thanks to Dennis Sweeney of Leigh who freely gives of his extensive rail history knowledge. Mike Taylor for his colour images of locos around Walkden. Thanks to Nick Busschau and Dave Ingham for their fine colour images of *Repulse*.

Thanks to Roger Fielding for the use of his images, also Charlie Verrall for his photos, Martyn Hearson for his photo of NSR No. 2 at Chatterley Whitfield. Thanks to the enthusiastic John Philips for allowing me to use his Walkden and Linnyshaw area images. Thanks especially to Steve Leyland of Bolton for allowing access to images from his historically important photographic archive. Thanks to Anthony Coulls, Senior Curator at the National Railway Museum for information and photography relating to *Princess*, NSR/LMS No. 2. Thanks to Brian Wharmby for the use of his fascinating photographs. Thanks to Duncan McCormick, Local Studies Librarian at Salford Local Studies Library, for allowing use of the Newtown Colliery photograph. Thanks to John Taylor of Leigh for the use of his and his friend's aerial images.

Preface

The preceding volume *Walkden Yard* concentrated on the Yard, it's history and the development of the mineral railways. To have combined that research with the contents of this volume was not possible. This volume therefore documents the locomotives themselves which were either regularly maintained at Walkden Yard, based there, or had fleeting connections with the site. It is meant to be an illustrated reference catalogue of the locomotives and the collieries they served, the illustrations being all important and as much as possible being previously unpublished.

The old Bridgewater Trustees mineral railways were to become the Central Railways of the huge Manchester Collieries concern, formed in March 1929. The landscape with its changing, suddenly abrupt and often fierce gradients, was to be a cruel one for these colliery locomotives, which were worked virtually constantly to their limits. From Worsley to Linnyshaw Colliery, east of Walkden, the average gradient had been 1 in 52 with the occasional 1 in 30 stretch! In later days, a typical run might be from Astley Green Colliery, south west of Walkden, on the 30m (98ft) contour to Mosley Common Colliery at 54m (177ft), Walkden Yard at 89m (291ft), Ashtons Field at 110m (360ft), with the branch to Brackley Colliery ending up at 114m (374ft).

To summarise the system (courtesy of The Industrial Railway Society): from the west we have BR at Astley Moss Sidings heading north to Astley Green Colliery and loco shed (1¾ miles). The line headed east to Boothsbank Tip (otherwise known locally as Boothstown Basin) canal wharf (3 miles) where coal was tipped into barges. A northerly change of direction brought the line to the important Mosley Common Colliery (3½ miles), beyond which, connection was made with BR west of Ellenbrook Station. Continuing north-east to Walkden led to connections with two BR lines; Ellesmere Sidings ½ mile west of Walkden High Level Station, and Walkden Sidings ¼ mile west of Walkden Low Level Station. Next stop was Walkden Yard (4¾ miles) with loco shed. North-west of Walkden Yard was Ashtons Field Colliery (5¾ miles), with coal blending plant and landsale depot (Buckley Lane Depot).

From Ashtons Field a line ran east for ¾ mile to a junction. A branch turned north-east to a connection at Linnyshaw Moss (¼ mile) with the BR Kearsley branch. Beyond this point the line from Ashtons Field headed south-east, later south to Sandhole (Bridgewater)

Colliery with its washery and loco shed (2 miles). South of Sandhole led to a BR connection at Sandersons Sidings, ½ mile north-west of Worsley Station. The line then continued to Worsley Tip canal wharf (3½ miles). North of Sandhole Colliery heading north-east then east was a line to the Moss Lane Landsale Depot, Pendlebury (2 miles).

From Ashtons Field Colliery, a line ran north-west to a junction with a line north-east to Dixon Green (Farnworth) Landsale Depot (½ mile). The line carried on west to Brackley Colliery and loco shed (1¾ miles) and the ever-growing Cutacre waste tips (2 miles) which have been worked and landscaped as I write. A connection with the BR Little Hulton Branch was finally made (2¼ miles).

As the older collieries gradually closed, the Moss Lane branch closed in 1956, Sandersons Sidings to Worsley Canal closed around 1961, the Sandhole Colliery to Sandersons Sidings line closed in 1966. The Dixon Green branch line closed in 1966, the Ashtons Field Colliery to Brackley Colliery closed in February 1968. The Linnyshaw Moss to Sandhole Colliery line closed in September 1968. The last lines from Ashtons Field Colliery to Linnyshaw Moss and the Astley Moss sidings closed in October 1970.

The staff at Walkden Yard worked wonders constantly overhauling these hard working locomotives. They had the skills to tackle virtually any job required with the benefit of generations of experience handed down, many times from father to son.

The images included in this study are mainly of locos which passed through Walkden Yard for maintenance or eventual scrapping. A collection of photographs had been built up at Walkden Yard over the years, held in the manager's office. It was originally brought to my attention in the early 1990s by Joe Cunliffe, the fifth Master of the Yard/Manager/ Superintendent in Walkden Yard's history. As well as official works photographs from the days of Manchester Collieries after 1929, Joe rescued photos taken in the days of Lancashire Associated Collieries and the NCB. He appears also to have been given photographs by enthusiasts he allowed to visit the site.

I realised virtually all of the images had not been published before, which is unusual in today's enthusiastic world for all things railways and locomotives. I felt a publication based on locos which had some connection with Walkden Yard, however fleeting, would be worthwhile.

Additional unpublished images held in private collections also surfaced during research and were added to the study. Not all the locomotives which passed through Walkden Yard are featured and certain of the photographs were taken away from the Yard.

It has also been made clear to me that there are dangers in trying to document the movements of the various locomotives over their lives and that there are occasional conflicts among rail historians in this respect. To try and resolve these would be impossible due to the lack of an archive record documenting all the movements of the locomotives.

The images, I feel, are worthy of publication even though they are mainly of static locos and at times not of the best quality, along with additional historical information which will be new to many readers. Sadly, very few of the images had the photographers' details with them so I can only apologise if you spot an image taken by yourself or which you now claim copyright by ownership which I may have thought to be of great age!

Alan Davies
Tyldesley
2014

The locomotives

The entries are arranged in alphabetical order with a chapter for each locomotive for ease of retrieval for the reader, rather than one of various other sequences such as by maker, date of construction, date of photograph, owner at the time of the photograph, etc. Not all locomotives which ever entered Walkden Yard or worked the Central Railways system are listed due to the lack of archives and photographs. Other locos are included purely because they were maintained at Walkden once or twice as the coalfield passed through various reorganisations. One is included because Walkden men were called out to carry out repairs – there is always some form of Walkden input.

The surviving maintenance records once held at Walkden Yard are very detailed but only for specific time periods. They are fairly similar as regards the work carried out at major overhauls so rather than repeat the information for every locomotive I thought it better to give a fairly detailed example of one or two locomotives' maintenance records with date summaries and selected excerpts for the rest.

Alison

Newly overhauled and painted, loco *Alison* (Hunslet Engine Co. Ltd 3163/1944) photographed at Walkden Yard in 1970. The Yorkshire Engine Co. 200hp diesel-electric 0-6-0 of 1957 (works number 2660, ex-Agecroft Colliery and Ashtons Field) is preparing to take the loco to Astley Green Colliery for delivery to Gresford Colliery, North Wales, by low loader.

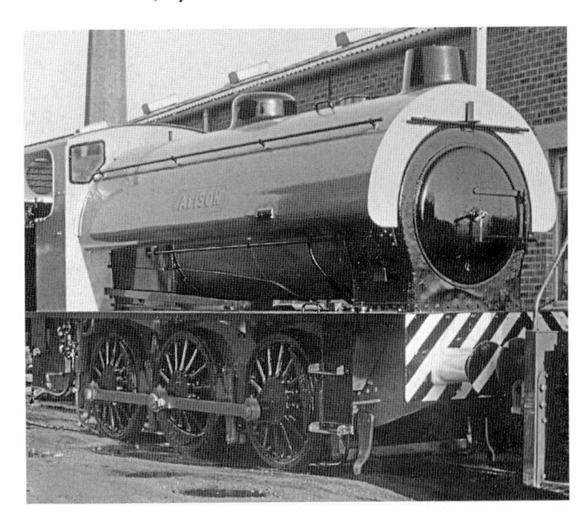

A closer view of *Alison* gleaming in Walkden Yard awaiting transfer to Gresford Colliery in November 1970. Note the fibreglass cowl chimney surround fitted during the Hunslet underfeed stoker upgrade, nicknamed the 'pork pie'. Within this was a chimney of flowerpot shape, much derided by all who saw it!

Loco *Alison* soon after delivery to Gresford Colliery, North Wales, in November 1970. Probably shunting pit waste in the ramshackle mixture of steel and wooden sided wagons. The colliery closed on 10 November 1973, *Alison* then moving to Bold Colliery, St Helens. At the time of writing *Alison* is running on the East Lancs Railway as WD132 *Sapper*.

A fine view of newly overhauled and dazzle painted *Alison* alongside Walkden Yard around August–September 1970, loco shed in the distance. Its Hunslet upgrade had comprised a mechanical stoker, the introduction of secondary air and steam into the firebox, and a Kylpor exhaust system. Note the three firebox secondary air holes. (Chris Tivey)

Proud Walkden Yard retiring manager Joseph Cunliffe climbs aboard *Alison* during its overhaul around October 1979. It received the name *Joseph* on 15 February 1980, returning to Bold Colliery, St Helens, where production ceased in November 1985. Note the standard chimney is back in place.

Loco *Alison* was constructed in 1944 by the Hunslet Engine Co. Ltd of Leeds, works number 3163. The Hunslet Engine Company was founded in 1864 at Jack Lane, Hunslet, Leeds, by John Towlerton Leather, a civil engineering contractor, who appointed James Campbell (son of Alexander Campbell, a Leeds engineer) as his Works Manager.

The loco was originally supplied to the War Department, Bicester, and numbered WD 75113, later titled WD 132. Similar locomotives supplied to Walkden Yard during the war were among the most powerful locos in the No.1 Manchester Area of the newly formed National Coal Board in January 1947.

The classic Austerity design of the Hunslet Engine Company met wartime demand, being based on earlier Hunslet designs. The saddle tank, inside cylinder loco, comprised 51in driving wheels, a coupled wheelbase of 11ft, 18in x 26in cylinders, a working steam pressure of 170 Ibs per square inch, a healthy 75 per cent pressure tractive effort of 21,060 lbs, and a working weight of 48 tons 4 cwt. Height of the loco, rail to chimney was 12ft 1in, width 8ft 2 1/2in, length 30ft 4in.

At some point after delivery it underwent repair by the GWR, Oxford. It moved afterwards to the Cainryan Military Railway, Stranraer, in 1950, to Longtown in 1952, and to Bramley Central Ammunition Depot, Hampshire, by 1955. Returned to Bicester for storage in 1957, being renamed *Sapper* from 1959 to 1961. Sold back to the Hunslet Engine Co. Ltd in October 1962. Rebuilt in 1964–1965 by Hunslet, it received a new works number of 3885. Loaned to the Homefire smokeless fuel plant adjacent to NCB Coventry Colliery on 1 May 1965, returning to Hunslet on 24 May 1966.

Purchased by the National Coal Board in July 1970, it was overhauled at Walkden Yard and dispatched to Gresford Colliery, near Wrexham, in November 1970, to be named *Alison*. Gresford Colliery closed on 10 November 1973 due to exhaustion of reserves and poor geology. The loco moved to Bold Colliery, St Helens, in April 1974, but was not put into use until October of that year. It went back to Walkden Yard to be overhauled *c.* October 1979 onwards, receiving the name *Joseph* on 15 February 1980 (after the retiring Joseph Cunliffe, manager Walkden Yard), returning to Bold Colliery where production ceased in November 1985.

Loco WD 75113/WD 132/*Sapper*/*Alison*/*Joseph* headed off in April 1992 to the South Devon Railway, Buckfastleigh, re-emerging once more as *Sapper*. By September 1992 it was on loan to the Swanage Railway, Dorset, returning to the South Devon in April 1994.

It was sold in 2009, returning to Lancashire at last, to the East Lancashire Railway as WD132 *Sapper* where at the time of writing (2013) it was operational. Many photographs of the loco in its WD splendour can be seen via Flickr on the internet, never mind a visit to Bury to see it actually in action.

Allen

A possibly late 1950s to early 1960s shot of Hudswell Clarke (1777/1944) WD 71500 Austerity *Allen* at Walkden Yard. The loco arrived at Walkden Yard on hire on 19 April 1945. It was bought in July 1946 and assigned the name *Allen*. Always used on the Central Railways or at Astley Green Colliery, it was scrapped in September 1968.

Loco Allen.
Came in Loco Shop Walkden yard
Monday 23-7-56
Engine lifted wheels taken out, and tyres
turn up to standard gauge

6 axle box brasses renewed, 6 axle box keeps built-up
and refitted
6 axle box keep pins renewed, 6 armstrong oiling pads
renewed.
2 pair of Big end brasses renewed
piston taken out. 4 piston rings renewed
slide bars lined up with cylinders. 4 slide blocks
remetalled, machined and fitted to slide bars.
6 side rod bushes renewed
2 side rod coupling bushes renewed
6 crank pins collars welded up machined to size
and 6 new taper pins fitted
6 Brake hanger brackets ground up.
3 Brake beams welded up at each end. and turned up
18 Bushes renewed in brake hangers.
6 New brake block pins turned and fitted
brake shaft taken down, and turned up at each end.
Brake shaft brackets taken down, bored out, and
Bushed to suit shaft. Pistons and Valve spindle
packings renewed 4 new sand jets fitted regulator
handle welded up and refitted
6 Bearing springs overhauled. All boiler mountings
overhauled. Floorboards renewed in cab
Front & Back drawbars and springs renewed.
cylinder drain taps ground in
Completed 24.8.56
Returned to Astley Green 27.8.56. ABaillie Thomason.Co.

The Walkden loco shop record for *Allen* coming in for repairs on 23 July 1956. Not a total boiler and all job, mainly bushes, bearings and worn parts being renewed and back to Astley Green Colliery. One month turnaround.

Allen, *Edward* and *Francis* line up alongside the Ellesmere Colliery banking at Walkden Yard, viewed south on 27 March 1968. By September 1968, *Allen* had been scrapped at Astley Green Colliery, *Francis* at Walkden in October 1968 and *Edward* at Walkden in October 1969. (Philip Hindley)

Named (in July 1946) after Allen Beaumont, Mining Agent for Manchester Collieries Eastern District by 1946. This Hudswell Clarke (1777/1944, WD 71500) Austerity loco was identical in specifications to *Alison/Sapper* described earlier. Hudswell Clarke was founded as Hudswell & Clarke in 1860 in Jack Lane, Hunslet, Leeds (the base four years later of the Hunslet Engine Co.) In 1870, the name was changed to Hudswell, Clarke & Rogers. There was another change in 1881 to Hudswell, Clarke & Company. The firm became a limited company in 1899.

Although the Hunslet Engine Company were responsible for the Austerity design, they subcontracted some construction work to Andrew Barclay Sons & Co., W. G. Bagnall, Hudswell Clarke, Robert Stephenson and Hawthorns and the Vulcan Foundry, Newton Le Willows.

The loco left the works on 16 December 1944. It arrived at Walkden Yard on hire to Manchester Collieries on 19 April 1945, being set to work five days later. Wartime livery was retained until it was bought in July 1946, when it was painted in black with red lining and assigned the name *Allen*. On nationalisation in January 1947, it was to be used by the NCB solely on the Central Railways or at Astley Green Colliery, being based at Walkden Yard. In a summary of locos in No. 1 Manchester Area of February 1947 it was commented alongside *Allen*'s entry and others adjacent: '*These Austerities are doing excellent work.*'

On 12 May 1950, it was transferred to Astley Green Colliery. *Allen* found itself in Walkden Yard for repairs in 1950, then 8/1/1951 to 30/5/1951, 23/5/1953 to 23/3/1954, 1956, 2/1960 to 6/1960, 1/1963 to 3/1963, early 1965, mid-1966 and mid-1967. It was scrapped at Astley Green in September 1968.

Amazon

Austerity *Amazon*, probably photographed at Haydock coal yard in the mid-1960s. Into Walkden Yard for repairs in March 1963, returning to Haydock in July 1963. *Amazon* headed off to the Whitehaven coalfield in October 1967 and was, along with many other locos, scrapped at the former Ladysmith coal washery *c*. September 1976.

The nameplate on *Amazon* was originally fitted to the first of the six 0-6-0 well tanks designed by Josiah Evans, constructed at Haydock between 1868 and 1887, the earlier style of lettering evident. The first *Amazon* was scrapped in 1935. The later *Amazon* was built to the standard Austerity design at the Vulcan Foundry, Newton le Willows, Lancashire (5297/1945 WD 75307), the loco was despatched to the Longmoor Military Railway in August 1945. Post-Second World War it returned close to where it had been constructed, being sold to Richard Evans & Co. Ltd, Haydock Collieries, in April 1946. On nationalisation in January 1947, the NCB utilised the loco in the St Helens Area until approximately October 1967, when it was transferred to the Cumberland Coalfield, in particular Harrington No. 10 Pit, west Lowca, north of Whitehaven. Although not working on the Central Railways, this photograph was among the Walkden Yard collection due to the loco undergoing repairs there, being sent from Haydock in March 1963, returning there in July 1963. She was scrapped at the former Ladysmith Colliery site, Whitehaven, which had been retained as a coal washery (closed 1975) around September 1976.

Atlas

Atlas (Peckett 1177/1909) seen probably at Bedford Colliery, Leigh, around the early 1960s. An ex-Fletcher Burrows & Co. Ltd Atherton Collieries locomotive purchased during the company's major period of investment. *Atlas* worked at Chanters, Gibfield and Howe Bridge Collieries, Atherton, as well as Gin Pit, Tyldesley and Bedford Colliery, Leigh.

Atlas seen on 28 January 1962 awaiting overhaul at Walkden Yard. Returned to Bedford Colliery, Leigh, in December 1962, *Atlas* remained there for three more years until heading off to Walkden Yard in November 1966 for boiler repairs (not carried out). (P. Eckersley, Brian Wharmby)

Atlas was taken out of service from Bedford Colliery, Leigh, being sent to Walkden Yard in November 1966 where this photograph was taken, eventually scrapped in October 1968. The foreground axles with deep balance weights appear to be those from either locomotive *Joseph* or *Bridgewater* (Hunslet 1456/1924, 1475/1924) whose remains languished at Walkden for many years until late 1968.

This internal cylinder 0-6-0 saddle tank (1177/1909) was built by Peckett & Sons Ltd (est. 1864) at their Atlas Works in Bristol. With a coupled wheelbase of 11ft 9in, 46in driving wheels and cylinders of 16in x 22in, a precisely stated tractive effort of 14,692 lbs was specified at 75 per cent of the 160 lbs working pressure, working weight being 40 tons.

Purchased new by Fletcher Burrows & Co. Ltd, owners of the Atherton Collieries, during their period of major investment 1904–13. Initially the loco worked on the LNWR lines from Gibfield Colliery west of Atherton and Howe Bridge Colliery to the south-west, taking poor grade 'slack' coal to the washery at Chanters Colliery to the east of Atherton.

Atlas was based at Howe Bridge Colliery by August 1938, moving to Chanters colliery by March 1946. Moving from there a mile to Gin Pit, Tyldesley, in 1946, possibly for repairs or overhaul, then back to Chanters in late 1946. The loco's working district came under the new No. 2 (Wigan) Area after 1 January 1952. *Atlas* would now be overhauled at Kirkless Workshops, Wigan or Parsonage Colliery, Leigh. Soon after, in 1953, the loco moved for a few months to Bedford Colliery, Leigh, then returned to Howe Bridge Colliery and Gibfield Colliery, Atherton. Following major boiler repairs in 1956 it operated around Chanters Colliery, Atherton, until 1958. A transfer to Bedford Colliery then took place. 1961 saw further area reorganisation with the new East Lancashire Area being created, Walkden Yard now taking over major overhaul work on all the locomotives in the North West Division including North Wales.

Atlas was overhauled at Walkden Yard in April 1962, returning in December of that year. The loco remained at work at Bedford Colliery for three more years until heading off to Walkden Yard in November 1966 for boiler repairs. These repairs were not to be carried out, the loco then awaited its end which came in late 1968.

Bedford

Bedford laid aside in Walkden Yard *c.* late 1930s. This venerable loco dated back to 1865, built by Manning Wardle at their Atlas Works Bristol (151 or 165/1865). Note the dome lagging exposed and very worn wheel tyres. Enthusiasts monitoring Walkden pinned down the date of final scrapping to late 1950.

This venerable internal cylinder 0-6-0 side tank loco dates back to 1865 and was built by Manning Wardle at their Atlas Works, Bristol (151 or 165/1865). The company had been established in 1859 by John Manning, C. W. Wardle and A. Campbell at the Boyne Engine Works, Leeds.

With a coupled wheelbase of 14ft, 46 ½in driving wheels and cylinders of 15 ¼in x 22 ½in, a tractive effort at 75 per cent working boiler pressure of 13,300 lbs was specified. Working steam pressure was 140 lbs per sq. in, working weight 30 ½ tons.

Bedford was originally built for the Potteries, Shrewsbury & North Wales Railway. The most accurate date for the purchase of the loco by John Speakman & Sons Ltd, Bedford Colliery, Leigh, is 1910. After the formation of Manchester Collieries in March 1929 it was to be transferred in 1933–34 to Astley Green Colliery. Into Walkden Yard for repairs in October 1938, November 1939 and from March to August of 1940. Sent back to Astley Green Colliery under the NCB in January 1947. Remarks by the NCB in their February 1947 survey of locomotives in No. 1 Manchester Area were: 'Out of service. Extensive repairs required'. NCB additional notes added to the sheet of 1962 state the loco was scrapped in 1947, although railway enthusiasts regularly monitoring activities at Walkden have pinned down the date of scrapping to late 1950.

Black Diamond

Ex (1922) Clifton & Kersley Coal Co. Ltd, Spindle Point Colliery, Farnworth, loco Sharp Stewart *Black Diamond* (2742/1878) and ex (post 1929) Bridgewater Collieries Manning Wardle *Stanley* (366/1871) at the Astley Green Colliery sheds in August 1938. *Black Diamond* was scrapped in October 1949. (Late Alex Appleton)

An outside cylinder 0-4-0 saddle tank, built in 1878 by Sharp Stewart & Co. Ltd, established 1843 in Manchester (2742/1878). The company moved to Glasgow in 1888. With a coupled wheelbase of 7ft 8in, 48in driving wheels and cylinders of 15in x 20in, its tractive effort (at 75 per cent of the working boiler pressure of 140lbs) was 13,236 lbs. The working weight of the loco was 28 tons.

The loco was internally transferred in 1922 from Spindle Point Colliery, east of Farnworth (closed 1928), to the relatively new major undertaking of Astley Green Colliery. The pit was sunk from 1908 to 1912, managed by the Pilkington Colliery Co. Ltd, a subsidiary of Clifton & Kersley Coal Co. Ltd.

Remarks by the NCB in their February 1947 survey of locomotives in No. 1 Manchester Area were: 'Shunting at Astley Green sidings. In fair condition.' NCB additional notes added to the sheet dating to 1962 state the loco was scrapped in 1949, the actual month being October of that year.

Bold

Bold (0-4-0ST Peckett, 1737/1927) arrives at Walkden Yard on 1 August 1965 for repair. Normally the NCB Haydock workshops would have dealt with this but they had closed in March 1963.

A formidable Pickford's Scammell haulage wagon delivers *Bold* to Walkden Yard, 1 August 1965, seen ready for unloading onto the loco shop's lines. By this time Pickford's were part of the National Freight Corporation (NFC), owned by the Treasury.

Bold (0-4-0ST Peckett, 1737/1927) was a temporary visitor by Pickford's low loader to Walkden Yard on 1 August 1965 due to the closure of NCB Haydock workshops in March 1963. The loco was ex-Sutton Heath & Lea Green Collieries Ltd (Bold Colliery, Lea Green Colliery and Sherdley Colliery) on nationalisation of the coal industry on 1 January 1947. Major overhauls of locos at Haydock Workshops had finished by late 1956. *Bold* is known to have gone into Haydock workshops for minor repairs after September 1959 while working at Bold Colliery, afterwards (December 1960) being sent to Ravenhead Colliery, central St Helens. By March 1963, Haydock workshops had closed hence Walkden being the only option for repair in 1965. *Bold* returned to Ravenhead Colliery in January 1967

Brackley

0-6-0 *Brackley* (Naysmith, Wilson 912/1910) in works photography 'colours' on a company product card. The loco cost the 2013 equivalent of £187,527. Tractive effort of 22,300 lbs, weight 56 ½ tons, supplied with wooden brake blocks. Identical loco *Ellesmere* was purchased by Bridgewater Collieries two years later.

GAUGE OF RAILWAY .... FT. .... INS.

CYLINDERS .... DIA.

,, STROKE ....

WHEELS.—COUPLED .... DIA.

,, ... ,,

WHEEL BASE—FIXED ....

,, ,, TOTAL ....

HEATING SURFACE OF FIRE BOX.—SQR. FT.

,, ,, TUBES— ,,

,, ,, TOTAL ,,

GRATE AREA .... SQR. FT.

CAPACITY OF WATER TANKS...... GALLONS

,, BUNKER .... CUB. FEET

WEIGHT IN WORKING ORDER......... TONS

,, EMPTY.......................... ,,

Reverse of the works product card for *Brackley* with standard specification fields awaiting completion. The ratio of heating surface to grate area gave an indication to the buying engineer of the potential efficiency of the design, and many other factors affecting this.

A fine view of *Brackley* at Walkden Yard around the 1930s. Based mainly at Walkden Yard and Sandhole Colliery during its working life. Taken out of service by April 1956 and cut up in February 1957.

This large, internal cylinder 0-6-0 saddle tank (912/1910) was built by Nasmyth, Wilson & Co. (est, 1867) at their Bridgewater Foundry, Patricroft, Manchester, in 1910. The Heavy Expenditure books at the Lancashire Record Office state that *Brackley* was purchased by Bridgewater Collieries for £2002 (or £187,527 in 2013 money).

The loco had a coupled wheelbase of 15ft 0in with 51in driving wheels and cylinders of 18in x 26in. Its 10ft 6in long by 4ft 6in diameter boiler consisted of 901 sq. ft of tube surface, an 82 sq. ft firebox and a 15.5 sq. ft grate. A tractive effort of 22,300 lbs was specified at 75 per cent of the working boiler pressure of 180 lbs. Water capacity was 1,750 gallons, coal capacity 1 ¾ tons. Weight empty was 42 tons 8cwt, working weight 56 ½ tons. The loco was supplied with wooden brake blocks.

Brackley was the first locomotive to be built in Nasmyth, Wilson's new construction shop in 1910. The identical loco *Ellesmere* was delivered to Bridgewater Collieries two years later.

A life in the shops

As an example of the regularity of visits to the loco workshops at Walkden Yard by *Brackley* and other locos, apart from its annual boiler inspection/yearly overhaul in August/September/October of each year, the following jobs were carried out from 1922 to 1953. Looking at the records much other minor work was also carried out, such as renewing packings, bearings, reaming out, renewing floorboards, brake blocks, etc., which I have left out. Records have not survived for the locos' last three years;

1922 New boiler

1925 New cylinders.

August 1931 New axle boxes, wheels turned up, new piston rings, new valves, new buffer springs and other work.

May 1933 New piston rings, new slide valves, sanding gear overhauled, crank pins filed up, wheels turned up, valves reset, 62 copper tubes fitted in boiler, smoke box bottom cemented, and other work.

October 1934 Wheels re-tyred, all motion overhauled, new axle box brasses, copper patch around fire hole ring, new fire doors, new fire bars, and other work.

November 1935 Boiler lifted, wheels turned up, axle boxes lined up, two new buffers, new slide valves, 338 copper stays renewed, rivets at firebox end renewed, and other work.

March 1937 Wheels turned up, all motion overhauled, slide valve renewed, new piston rings.

June 1938 Fame lifted, wheels out, new tyres fitted, new axle box brasses, new piston rings, all boiler mountings taken off and repaired, leading and trailing drawbars renewed. Smoke box tube plate condemned by Insurance Company. Boiler sent away to Houghs Ltd (boilermakers) Newtown, Wigan. Fitted with 165 new copper tubes. Other general work carried out. Engine back in work 17/11/1938

November 1939 Driving axle broken near RH axle box, axle renewed, other work carried out.

August 1941 Wheels retyred and turned up, R & LH driving axle box brasses renewed, bearing springs repaired, valves reset, boiler mountings overhauled and other work.

November 1941 (twice) 8 second hand tubes fitted, R & LH trailing springs, pillars and brackets renewed, 8 more second hand tubes fitted.

May 1942 4 copper stays renewed.

November 1942 6 new copper tubes.
February 1943 2 new copper tubes.
April 1943 Boiler removed, wheels off and trued up, axle boxes and brasses renewed, new piston rings, brake work renewed, all boiler mountings renewed, sanding gear overhauled, tank and bunker repaired. Boiler copper tubes renewed, 277 copper stays renewed, 30 smoke box rivets renewed, outside corners of firebox built up by electric welding.
July 1944 Axle box brasses renewed, lubricators overhauled, slide valves renewed.
October 1944 Boiler inspection, 4 broken screw stays renewed.
January 1945 Side rod bushes renewed.
February 1945 Engine lifted, wheels removed, tyres turned up, six new drivers, axle box brasses renewed, piston rings renewed, brake work overhauled, front buffers repaired, chimney renewed, bunker repaired and other work.
August 1945 Engine lifted, RH axle box brasses renewed.
January 1946 RH piston head and rings renewed, RH inside cylinder cover renewed, slide bars lined up.
November 1946 Boiler removed, being stripped for internal examination. New back plate fitted, new firebox, new set of copper tubes, stays and brackets, 18 new longitudinal stay bolts fitted.
April to October 1947 Engine lifted, wheels out, tyres renewed and turned up to suit horn blocks, axle box brasses renewed, crank pins renewed, piston heads and rods renewed, boiler mountings overhauled, R&LH injectors repaired, frame repaired, cylinder lubricators and axle boxes repaired, cab and tank repaired, and other work.
January 1948 Fusible plugs renewed.
June 1948 Side rod bushes renewed.
August 1948 Annual boiler inspection, in working order.
Feb 1949, LH Piston ring renewed, slide valves reset, water gauge taps repaired.
September 1949 Annual boiler inspection, most of the tubes in the lower section of the fire box expanded and re-ferruled, lap crack caulked and other work.
January 1950 Fusible plugs renewed, fusible plugs renewed, slide valves renewed
September 1950 Annual boiler inspection, packings renewed, blast pipe renewed, side rod bushes renewed, tank repaired and other work.
February 1951 RH driving axle box broken, tanks renewed, wheels off, turned up, axle boxes repaired, bearing springs and pins renewed, piston rings renewed, new slide valves, valve motion overhauled, other general rebushing and repacking work, engine painted, engine tested May 1951.

Further repairs are known to have been carried out at Walkden in late 1953.

Thought of by some observers as a reserve engine, rather than one for full time demanding work, its service record telling perhaps another story, *Brackley* had mainly worked between Walkden Yard and Sandhole (Bridgewater) Colliery, where it was based for many years.

Remarks by the NCB in their February 1947 survey of locomotives in No. 1 Manchester Area were: 'Suitable for duty. Major repairs in hand.' Those repairs were indeed to be in hand, taking from April to October 1947. Additional notes added to the sheet in 1962 state the loco was scrapped in 1956, although enthusiasts monitoring Walkden Yard remarked that the loco was taken out of service by April 1956 then cut up in February 1957 by Stone & Hutchinson.

Bridgewater

Company image of 0-6-0 side tank *Bridgewater* (1475/1924) built by the Hunslet Engine Co. Ltd, Jack Lane, Hunslet, Leeds. Note for an industrial loco the Walschaert's valve gear and also the Ross 'pop' safety valves, dome mounted. Tractive effort 20,700 lbs, weight 50 tons 3 cwt.

A view *c.* 1930s of *Bridgewater* probably at Brackley Colliery, east of Little Hulton. After Manchester Collieries' formation in 1929, *Bridgewater* could be found at Brackley with identical loco *Joseph*. Their run was from Brackley Colliery to Ashtons Field Colliery, to Dixon Green (Farnworth) coal landsale yards. The growing coal tips at Cutacre, south of Brackley, were also served.

Bridgewater in for repairs on 24 March 1957 at Walkden, looking well used and in need of a paint job. Seen sporting its original chimney, soon to be removed. (P. Eckersley, Brian Wharmby)

Bridgewater seen at Walkden on 26 June 1960, probably for minor repairs. Bridgewater Offices in the distance. New chimney in place, damage to the cylinder cover sheet evident. (P. Eckersley, Brian Wharmby)

Bridgewater mid-1960s at Walkden Yard. Smoke reduction steam jet tubing is visible along the top of the boiler to the chimney-top ring, also the three entries to the pipes carrying secondary air back to the firebox high up on the smokebox front (three more on the other side not visible).

This purposeful outside cylinder 0-6-0 side tank (1475/1924) was built by the Hunslet Engine Co. Ltd of Jack Lane, Hunslet, Leeds (est. 1864). The hive of activity which was Jack Lane was also the home from 1860 of locomotive manufacturers Hudswell Clarke. *Bridgewater* was ordered in November 1923, leaving the workshops on 18 June 1924, painted olive green and lined in yellow and vermillion. The identical *Joseph* was ordered at the same time, costing Bridgewater Collieries £3,060 whereas *Bridgewater* cost £3,160 (£150,289.60 in 2013).

Unusually for an industrial locomotive, use was made of Walschaerts valve gear which enabled the driver to fine tune the steam engine's operation in a continuous range of settings from maximum economy to maximum power, useful on the punishing Bridgewater system. The two Ross 'pop' safety valves and whistle being dome mounted was also an interesting feature on the locomotives.

With a coupled wheelbase of 13ft 0in, 45in driving wheels and cylinders of 18in x 22in, a tractive effort of 20,700 lbs was specified at 75 per cent or 160 lbs boiler pressure. A maximum of 23,500 lbs was also specified at the insured maximum boiler pressure of 180 lbs, the working loaded weight being 50 tons 3 cwt. Overall width was 8ft 8in, maximum axle load 17 tons 8cwt, tube heating surface 891 sq. ft, firebox heating surface 100 sq. ft, grate area 17.87 sq. ft, tank capacity 1,100 gallons, bunker capacity 2 tons.

Remarks by the NCB in their February 1947 survey of locomotives in No. 1 Manchester Area were: 'In fair condition. To be fitted shortly with new firebox and cylinders which are in stock.'

Loco."Bridgewater": The Chairman explained that in anticipation of the approval of the Board, a second locomotive, to be known as "Bridgewater", had been ordered from the Hunslet Engine Co., Ltd., at a price of £3,160.

The purchase was confirmed, and the Agreement in connection therewith being produced, it was

RESOLVED that the Seal of the Company be affixed the thereto.

The board of Bridgewater Collieries decides to spend £3,160 (£150,289 in 2013) on the purchase of *Bridgewater* in late February 1924. (Lancs Rec Office NCB Bw 20/1)

After the formation of Manchester Collieries in March 1929, *Bridgewater* was usually to be found at Brackley Colliery, Little Hulton, along with *Joseph*. Their duty involved a run from Brackley Colliery to Ashtons Field Colliery, to Dixon Green (Farnworth) coal landsale yard and to the growing coal tips at Cutacre, south of Brackley Colliery. Journeys were also made to the LMS exchange sidings on the Little Hulton mineral line.

After nationalisation in 1947, *Bridgewater* worked generally on the Central Group railways, often on the main lines between Mosley Common Colliery and Sandhole (Bridgewater) Colliery. Shunting duties were undertaken at Sandersons Sidings en route to the canal tip at Worsley. The loco was sent to Astley Green Colliery on 21 May 1947 and was to be in use there until February 1948. Back into Walkden for repairs by mid-1948, it received a new boiler from the Hunslet Engine Co., returning to duties by early 1951. *Bridgewater* was to be seen at various locations on the system from then on, calling in at Walkden for further repairs in early 1953 and 1955. The newly arrived Austerity locomotives ousted her from her regular duties by 1956. In early 1957 it was back in Walkden again for repairs.

Interestingly, *Bridgewater* could complete its daily standard Mosley Common–Sandhole tasks in four hours. When the new Yorkshire Engine Company's 200hp 0-6-0 diesel electric loco arrived in 1957 the job took a full day. In the meantime *Bridgewater* was overhauled, receiving a new chimney, and given work in the Mosley Common area. In late 1959 the loco was sent to Astley Green Colliery. With the new East Lancashire Area colour scheme arriving in March 1962, *Bridgewater* would eventually be seen bedecked in the vivid maroon with yellow lining and NCB NW Division crest on the cab side, although it was to be 1966 in this loco's case. In the meantime Sandhole Colliery closed in September 1962.

The loco was back in Walkden for repairs in late 1963 to early 1964. From August 1964 until August 1967, it could be seen once more working at Astley Green Colliery apart from a spell of repairs at Walkden in mid-1965. Another overhaul took place in early 1966. Firebox problems led to removal from service and back into Walkden Yard in late 1967. It was taken out of service by March 1968 before having to suffer the indignity of a Giesl ejector conversion, finally being scrapped at Walkden Yard in October 1968.

Industrial historian and former engineer Geoff Hayes, writing in *Industrial Railway Record* No. 44 of 1972, stated that *Bridgewater* was a machine to be reckoned with, quickly setting up a reputation for unsurpassed weight shifting. Its operating characteristics, he said, were totally different from its shed mates for it boxed vigorously when pulling hard and quickly developed trailing axle box knock. Geoff commented that some drivers said it was not until after an overhaul and the knock had set in once more that the loco settled down! Geoff stated that the cab was quite commodious, the controls laid out for driving from the right hand side. The regulator handle was of the upright pattern but with an elegant curve to the right so that it could be operated by the driver when leaning out from the cabside. The steam brake valve was mounted over the top of the firebox, operated by a miniature inverted version of the regulator handle. No provision was made for operating the locomotive from the left hand side of the footplate. The whistle was operated by a lanyard from the cab. Gravity sanding was provided with large capacity sand boxes.

Carr

Loco *Carr* in October 1948 soon after arriving new at Astley Green Colliery (Hudswell Clarke 1812/1948). Named after J. F. Carr, Mining Agent NCB NW Div No. 1 (Manchester) Area. The loco arrived painted dark blue with yellow lining. (Late Alex Appleton)

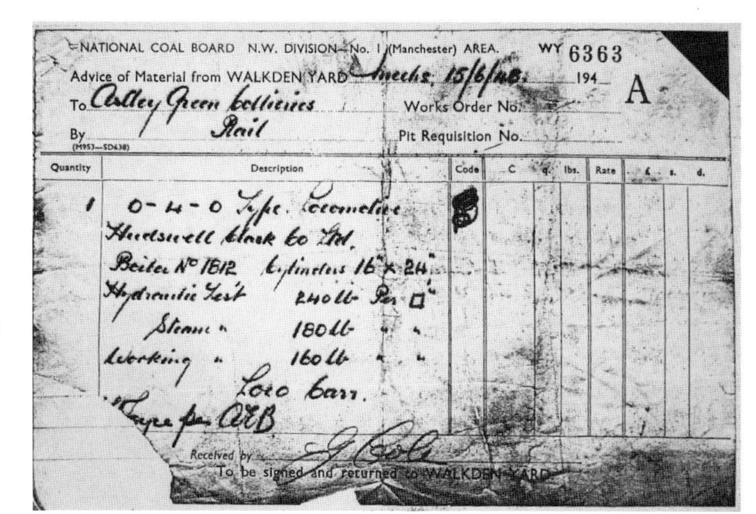

NATIONAL COAL BOARD N.W. DIVISION—No. 1 (Manchester) AREA. WY 6363

Advice of Material from WALKDEN YARD Mechs. 15/6/48. 194

A

To Astley Green Collieries Works Order No.

By Rail Pit Requisition No.

(M953—SD638)

Quantity	Description	Code	C	q.	lbs.	Rate	£	s.	d.
1	0-4-0 Type. Locomotive								
	Hudswell Clarke Co Ltd.								
	Boiler No 1612 Cylinders 16" x 24"								
	Hydraulic Test 240lb Per □"								
	Steam " 180lb " "								
	Working " 160lb " "								
	Loco Carr.								

per AEB

Received by

To be signed and returned to WALKDEN YARD

Just to make sure there is some documentation in case it goes missing! Loco *Carr* is despatched new to Astley Green Colliery on 15 June 1948 from Walkden Yard, the advice note duly signed at Astley and returned.

Carr at Astley Green Colliery sometime in the late 1940s or early 1950s. Steam tested to 180 lbs, working pressure 160 lbs. Coupled wheelbase of 6ft 9in, 45in driving wheels and cylinders of 16in x 24in. Tractive effort of 16,400 lbs specified, working weight 36 tons.

Carr seen at Walkden for repairs on 14 January 1962, Walkden offices behind. (P. Eckersley, Brian Wharmby)

Loco *Carr* out of use at the rear of the loco shed at Astley Green Colliery on 24 August 1964. Withdrawn from Astley Green late 1965 and despatched to the death row of many colliery locos, Ladysmith coal washery, Whitehaven, surviving until November 1972. (Philip Hindley)

This loco was named after J. F. Carr, former manager at Brackley Colliery, Little Hulton, and in 1948 the Mining Agent, NCB No. 1 Group. In 1946, the Ministry of Supply (1939–1959) ordered four new and identical outside cylinder 0-4-0 saddle tanks from Hudswell Clarke Ltd, including loco *Carr* (1812/1948). The company was founded as Hudswell & Clarke in 1860 in Jack Lane, Hunslet, Leeds. In 1870, the name was changed to Hudswell, Clarke & Rogers.

The NCB records tell us this was boiler number 1812, hydraulically tested to 240 lbs, steam tested to 180 lbs, with a working pressure of 160 lbs. Having a coupled wheelbase of 6ft 9in, 45in driving wheels and cylinders of 16in x 24in, a tractive effort of 16,400 lbs was specified, working weight being 36 tons.

Remarks (added later than the other entries of February 1947, the loco being purchased on 4 June 1948) by the NCB in their February 1947 survey of locomotives in No. 1 Manchester Area were: 'Fitted with spark arrestor. Purchased new 1948.' The NCB received the loco on 4 June 1948, eleven days later sending it to Astley Green Colliery. *Carr* was in for repairs at Walkden from March to August 1955 and again in 1960. Based at Astley Green until late 1965, it was then transferred to the William Pit site, north Whitehaven (closed as a working colliery in 1954). It was scrapped at the former Ladysmith Colliery washery site in November 1972.

Charles

Charles c. 1964 at Walkden Yard sporting cabside NCB NW Div crest in use until 1967. Hunslet Austerity design, built by Hudswell Clarke (1778/1944), ex-War Department loco. With Giesl conversion chimney, and secondary air holes visible in the firebox. Walkden offices and distinctive clock tower visible behind. A Central Railways loco until August 1967, transferred to Ladysmith coal washery, Whitehaven, scrapped in August 1973.

Named after Charles Fish, who had replaced William Green as Manchester Collieries Transport Manager. This internal cylinder 0-6-0 saddle tank (1778/1944) ex-War Department loco was built to the Hunslet Engine Co. Austerity design by Hudswell Clarke of Leeds. *Charles* was the next loco ex-works after *Allen*, on hire to the Walkden system from December 1944 and being purchased post-war in July 1946. With a coupled wheelbase of 11ft 0in, 51in driving wheels and cylinders of 18in x 26in, a healthy tractive effort of 21,060 lbs was specified, its working weight being 48 tons. It was normally working on the Central Railways and was based at Ellesmere loco shed, Walkden, until 1967. Into Walkden for repairs in 1950, 1953, early 1954, 1956, early 1958, mid-1959, late 1963 to early 1964 (receiving the boiler off *Wasp* and tank from *Warspite*). Eventually transferred to the Whitehaven Coalfield, Ladysmith Colliery washery, in August 1967 and scrapped there in August 1973.

Crawford

A Wigan Coal & Iron Co. Ltd marketing lantern slide view of Sovereign Pit, north-west of Leigh, one of a series. In view alongside the screen sheds is 0-6-0 ST loco *Crawford*, built 1883 at the company's Kirkless works, Aspull. The loco worked initially at this colliery, seen here with its original short saddle tank. Sovereign Pit closed in December 1927.

The view looks old but is probably late 1950s to early 1960s at Gibfield Colliery, Atherton, White Lee farmhouse close to Wigan Road behind. *Crawford* by now has an enclosed cab and extended saddle tank. Coupled wheelbase 11ft 10 ½in, 51in driving wheels, cylinders 16in x 20in, working boiler pressure 125 lbs, tractive effort 9,440 lbs, working weight being 35 tons.

Locos *Violet* (0-6-0ST Nasmyth, Wilson 852/1908) and *Crawford* pass Howe Bridge village, with coal from Chanters Colliery to the east, in the late 1950s. Coal headed to Gibfield colliery and Bag Lane, Atherton sidings, off the former Bolton–Kenyon, later LMS line. *Crawford* was scrapped at Gibfield Colliery in May 1964. (Albert Leather)

This internal cylinder 0-6-0 saddle tank was built in 1883 by Wigan Coal & Iron Co. Ltd at their Kirkless, Aspull, works. With a coupled wheelbase of 11ft 10½in, 51in driving wheels and cylinders of 16in x 20in, a working boiler pressure of 125 lbs (insured to 150 lbs) produced a tractive effort of 9,440 lbs (11,330 at 150lbs), working weight being 35 tons.

Wigan-based consultant mining engineer Cornelius McLeod Percy had visited the LNWR works, Crewe, for WC&ICo. for a few months in 1864 to obtain designs for locomotives. After earlier designs with 14in x 20in cylinders, a standard WC&ICo. design of 0-6-0 with saddle tank and internal 16in x 20in cylinders was arrived at, *Crawford* being an example. The loco appears to have worked initially at the company's Sovereign Pit, Leigh, to the north of their Parsonage Colliery. Sovereign pit closed in December 1927.

Crawford joined the loco stock of the new Wigan Coal Corporation on its formation in May 1930 (the amalgamation of the Pearson and Knowles Coal & Iron Company Ltd, The Moss Hall Coal Company Ltd, the Wigan Junction Colliery Company Ltd and the Wigan Coal & Iron Company Ltd). The loco is known to have been based at Parsonage Colliery in 1932 and 1940, temporarily working at Chisnall Hall Colliery, Coppull, in 1943, returning to Parsonage Colliery.

The NCB locomotive survey of 1947 stated *Crawford* was in 'moderate condition', in use at Parsonage Colliery and 'suitable for duty'. Other examples of the WC&ICo. design surviving on coal industry nationalisation in January 1947 were: *Lindsay* of 1887 (restored by colliery manager and mining historian Donald Anderson and partners at his Quaker House Colliery, Winstanley, Wigan, from 1976 onwards, now static at Steamtown, Carnforth), *Balcarres* of 1892 and *Emperor* of 1893.

Crawford received the boiler and saddle tank from *Emperor* when it was scrapped in 1947. Maintenance records for *Crawford* do not appear to have survived apart from a single entry in a ledger at Walkden Yard:

> *Loco Crawford, at Parsonage [Colliery],*
> *Piston head, rod and rings renewed. Valve motion overhauled. 4/7/1949*

After another overhaul (at the NCB Kirkless Workshops, Wigan) *Crawford* was sent to Howe Bridge Colliery, Atherton, in March 1957. Howe Bridge Colliery closed in September 1959. The loco was sent to nearby Gibfield Colliery to the north, which ceased production on 30 August 1963. Taken out of use and described as 'derelict' by loco enthusiasts by January 1964, it was scrapped at Gibfield Colliery in May 1964.

Earlestown

Loco *Earlestown*, 0-6-0 ST IC Manning Wardle (1503/1900), paid a visit to Walkden Yard for repairs in 1957 as major overhauls of locos at Haydock Workshops had finished by late 1956. Luckily photographer John Philips was on site to record work in progress. *Earlestown* was scrapped at Haydock in June 1965.

Three locomotives were acquired by Richard Evans' Haydock Collieries in 1900. All were constructed by Manning & Wardle at their Jack Lane, Hunslet, Leeds, works. The company had been established in 1859 by John Manning, C. W. Wardle and A. Campbell at the Boyne Engine Works, Leeds. The company closed in 1927 after producing more than 2,000 steam locomotives.

The locomotives were given the names of locations adjacent to Haydock: *Garswood* 0-4-0 ST OC (1486/1900), *Earlestown* 0-6-0 ST IC (1503/1900) and *Newton* 0-6-0 ST IC (1504/1900). *Garswood* was scrapped in 1957 at Ravenhead Colliery, St Helens.

Earlestown began its working life with the long-established Richard Evans & Co. Ltd Haydock Collieries. It is known to have worked at Lea Green Colliery in 1951 and 1952, and Garswood Hall Colliery in 1959.

Major overhauls of locos at Haydock Workshops had finished by late 1956, hence the reason *Earlestown* paid a visit to Walkden Yard for repairs in 1957 (see photo). Lesser repairs to steam and diesel locos continued at Haydock until 1963. Some St Helens area locos were also sent for repair to Kirkless Workshops, Wigan. *Earlestown* was taken out of use by 1960, being scrapped in June 1965 at Haydock by Jose Holt Gordon Ltd of Chequerbent, north of Atherton. *Newton* was moved from Lea Green Colliery, St Helens, to Chequerbent, and scrapped there in August 1965

Edith

A photograph probably taken soon after loco *Edith* arrived at the Shakerley Collieries, Tyldesley, in 1887. This internal cylinder 0-6-0 saddle tank was built by the Hunslet Engine Co. Ltd, Leeds (436/1887). Coupled wheelbase 10ft 9 ½in, 36in driving wheels, cylinders 14in x 18in, working pressure 125 lbs, tractive effort of 9,200 lbs, working weight being 23 ½ tons.

In Manchester Collieries days, *Edith* was transferred to work between Howe Bridge Colliery and Gibfield Colliery, Atherton, in 1939. The 1947 NCB loco survey stated *Edith* was working at Gibfield Colliery, Atherton: *'In good condition. Suitable for duty.'* *Edith* was taken out of service at Howe Bridge Colliery in late 1956. It was to be May 1959 before it was finally scrapped. (Albert Leather)

Named after Edith Ramsden, daughter of William Ramsden, owner of the Shakerley Collieries, Tyldesley.

This internal cylinder 0-6-0 saddle tank was built (436/1887) by the Hunslet Engine Co. Ltd, Jack Lane, Hunslet, Leeds (est. 1864). With a coupled wheelbase of 10ft 9 ½in, 36in driving wheels and cylinders of 14in x 18in, a working pressure of 125 lbs, a tractive effort of 9,200 lbs was specified, working weight being 23 ½ tons. Purchased by William Ramsden & Sons Ltd for their Shakerley Collieries, Tyldesley (Nelson and Wellington pits). *Edith* was returned all the way back to Leeds for repairs in May 1905. Geoff Hayes, in his *Collieries & their Railways in the Manchester Coalfields*, interestingly states that loco *Edith* had side play incorporated into its leading and trailing axles and vertical pin joints in the coupling rods to allow for the negotiation of sharp curves present at these very old collieries.

After the formation of Manchester Collieries in March 1929, the directors' eyes turned towards the ageing and run-down Shakerley Collieries as a means of increasing their overall coal production allowance. The Shakerley concern had seen its most profitable days; one of its directors, W. H. Ramsden, had recently died (22 November 1934) leaving the equivalent today (2013) of £12,000,000. The company was purchased in March 1935, Wellington Colliery being closed only two months later, Nelson Colliery closing in October

1938. *Edith* was transferred to work between Howe Bridge Colliery and Gibfield Colliery, Atherton, in 1939.

The 1947 NCB loco survey stated that *Edith* was working at Gibfield Colliery, Atherton: 'In good condition. Suitable for duty.' *Edith* was taken out of service at Howe Bridge Colliery in late 1956. It was to be May 1959 before it was finally scrapped.

Edward

Loco *Edward* was built (3184/1916) by Hawthorn Leslie & Co. Ltd, Newcastle-on-Tyne, new to Clifton & Kersley Coal Co. Ltd. Seen here at Astley Green Colliery in the 1930s, Manchester Collieries wagons in tow. Astley Green was the loco's base for all its working life. Tractive effort 20,500 lbs, weight 48 tons.

Loco *Edward* and possibly the whole complement of transport staff at Astley Green Colliery, 1932. On the footplate, left to right: J. Grundy and R. Lawley. Left to right, standing: J. Hobson, J. Sharples, H. Bradshaw, H. Green, W. Edwards, J. Williams, H. Speakman, F. Pearson, R. Jones, D. Parkinson and A. Grundy. (Jimmy Jones)

A late 1940s to early 1950s view of *Edward* at Astley Green Colliery shed, viewed to the west, certainly no NCB paintwork or crests visible. A few knocks and dents now visible since the 1930s photograph.

Edward seen at Walkden on 24 April 1960, probably for minor repairs. (P. Eckersley, Brian Wharmby)

Edward at work at Astley Green Colliery, Monday 24 August 1964. Now resplendent (with imagination!) in the new maroon and yellow lining paint scheme and NCB NW Division cabside crest. *Edward* was scrapped at Walkden Yard in October 1969, a few months before Astley Green Colliery finally closed. (Philip Hindley)

Named after Edward Pilkington, a director of the Pilkington Colliery Co. Ltd (a subsidiary of The Clifton & Kersley Coal Co. Ltd), which was set up to manage the new Astley Green Colliery, sunk from 1908 to 1910. The outside cylinder 0-6-0 saddle tank was built (3184/1916) by Hawthorn Leslie & Co. Ltd, Newcastle-upon-Tyne, the shipbuilding and locomotive manufacturer. The original R. & W. Hawthorn was an early locomotive manufacturer, established in 1817. In 1885, shipbuilder A. Leslie Co. (of Hebburn-on-Tyne) and R. & W. Hawthorn merged to form the public company Hawthorn Leslie & Co.

Edward was supplied new to Clifton & Kersley Coal Co. Ltd in 1916. It was to spend all its working life at Astley Green Colliery and for nearly the whole life span of the pit itself. With a coupled wheelbase of 11ft 6in, 44in driving wheels and cylinders of 17in x 24in, a tractive effort of 20,500 lbs was specified at a working pressure of 170 lbs, working weight being 48 tons.

The Walkden Yard loco repair log records the arrival of *Edward* on 9 March 1931:

Boiler tubes replaced, wheels trued up, axle box brasses re-bedded on. Side rods bored out, new brasses fitted. All motion work overhauled. Completed April 7th 1931.

Few other repair and maintenance records have survived for *Edward*. It is known the loco was at Walkden Yard again for repairs from May to July 1938.

The NCB loco survey of 1947 stated that its present duties were: 'Astley Green shunting and Main Line'. It commented that minor general repairs were being carried out and that the loco was suitable for duty. *Edward* was scrapped at Walkden Yard in October 1969, a few months before Astley Green Colliery finally closed.

Ellesmere

The imposing *Ellesmere* at Walkden Yard in the late 1930s. Built (976/1912) by Nasmyth, Wilson & Co., Bridgewater Foundry, Patricroft, Manchester. The 42 ½ ton loco (56 ½ tons working) was to be found either at Sandhole Colliery or Walkden Yard sheds during its working life. Scrapped at Walkden in November 1963. (Glen Atkinson)

Named after Francis Charles Granville Egerton, 3rd Earl of Ellesmere (1847–1914) and not to be confused with loco *Ellesmere* of Fletcher Burrows & Co. Ltd, Atherton Collieries, which lost its name on formation of Manchester Collieries in March 1929, reverting back to the title '1861', the year of its manufacture by Hawthorns of Leith (244/1861).

This was an internal cylinder 0-6-0 saddle tank (976/1912) built by Nasmyth, Wilson & Co. (est. 1867) at their Bridgewater Foundry, Patricroft, Manchester, in 1912. The Bridgewater Collieries Heavy Expenditure books (NCB/Bw) at the Lancashire Record Office tell us that *Ellesmere* was purchased by Bridgewater Collieries in 1912 for £2,299 (or £233,941 in 2013). With a coupled wheelbase of 15ft 0in, 51in driving wheels and cylinders of 18in x 26in, a tractive effort of 22,300 lbs was specified for a working pressure of 180 lbs. The boiler measured 10ft 6in long by 4ft 6in, its heating surface made up of: tubes 901 square feet, firebox 82 square feet, total 983 square feet. The grate area amounted to 15.5 square feet. Water capacity came to 1,750 gallons, coal capacity 1.75 tons. The loco's empty weight was 42 tons 8 cwt, working weight 56 tons 10 cwt. Factor of adhesion was a healthy 5.7.

It is known that the loco was overhauled at Walkden Yard in 1922 including the fitting of a new firebox. New cylinders were fitted in 1928. Up to nationalisation in 1947 *Ellesmere* was to be found either at Sandhole Colliery or Walkden Yard sheds. The 1947 NCB loco survey stated that its present duty was: 'On Central Railways Walkden. Some shunting, but chiefly on long distance traffic between collieries, washeries and main line junctions.' It was also stated: 'Suitable for duty. In fair condition. Parts to hand for major repairs due.'

Ellesmere headed into Walkden Yard for repairs in late 1947, mid-1951 until early 1952, mid-1952, late 1953, mid-1954. Taken out of service, it was partly cut up by September 1956, the chassis remains being finally scrapped in November 1963.

Failsworth

This outside cylinder 0-6-0 saddle tank (2727/1907) was built by Hawthorn Leslie & Co. Ltd, Newcastle-upon-Tyne, the shipbuilding and locomotive manufacturer. The original R. & W. Hawthorn was an early locomotive manufacturer, established in 1817. In 1885 shipbuilder A. Leslie & Co. (Hebburn-on-Tyne) and R. & W. Hawthorn merged to form the public company Hawthorn Leslie & Co.

With a coupled wheelbase of 10ft 6in, 45in driving wheels and cylinders of 15in x 22 ½in, a stated tractive effort of 10,960 lbs was specified at 130lbs pressure, working weight being 30 tons. The NCB loco survey of 1947 stated that its present duties were shunting at Moston Colliery east of Manchester (owned by Platt Brothers & Co. (Holdings) Ltd). The survey also added that it was 'Adequate when in order' and 'Stopped for major repairs to boiler'.

After nationalisation in January 1947 the loco was eventually moved in February 1951 to Wheatsheaf Colliery, north of Swinton. According to former Walkden Yard Manager Joe Cunliffe in 1990 (died 1993), *Failsworth* was the first locomotive to be dealt with in the new loco workshops on 3 August 1953. It was overhauled once more at Walkden Yard in May to June 1955. It remained at Wheatsheaf Colliery until being scrapped in September 1958.

Francis

A fine photograph by that meticulous recorder of the industrial loco in Lancashire, Alex Appleton, of *Francis* (Kerr, Stuart, Stoke-on-Trent, 3068/1917) at Mosley Common Colliery, 21 May 1938. A local Eccles & District Co-operative Society Ltd wagon awaits its turn beneath the coal classification screens. (Alex Appleton)

1930s view of *Francis*, probably in the Mosley Common Colliery area where it was based at the time. Note the grimy driver's face coated with sweat and coal dust. (Glen Atkinson)

Francis viewed in the 1930s, most likely near Mosley Common Colliery. Coupled wheelbase of 12ft 0in, 48in driving wheels, cylinders 17in x 24in, tractive effort 19,652 lbs at 160 lbs, working weight 50 tons. Boiler heating surface 1,145 sq. ft, grate area 20 ½ sq. ft, 217 boiler tubes 1 ¾in diameter. (Glen Atkinson)

Francis surrounded by locos at Walkden shed in 1966, probably during the repair visit of 1965 onwards. *Francis* ceased working at Astley Green Colliery *c.* late 1967, being scrapped at Walkden in October 1968. (Roger Fielding)

Francis seen at Astley Green Colliery on 24 August 1964, internal wagon in tow. Note the smoke reduction steam jet supply pipe entering the chimney top, said to be an effective adaptation. (Philip Hindley)

This interesting shot taken at Walkden *c.* 1966 of *Francis* shows plainly how the loco had suffered the final phase of the Giesl ejector chimney conversion, marring its outline. Note the old steam jet pipe entering the smokebox above the handrail, Belpaire firebox and 'pop' safety valves above.

Francis north of Boothsbank coal tippler, 22 October 1965. The loco is facing north, standing in the entry to the two empties roads at the tippler. Wagons were hauled from Astley Green Colliery and pushed into the fulls sidings (those in front of the loco) then coupled onto empties for return to Astley Green. (Philip Hindley)

Named after Francis Egerton, 3rd Duke of Bridgewater (1736–1803), also known as The Canal Duke, the entrepreneur behind the construction of the canal to transport coal from his mines at Worsley to Manchester. The Bridgewater Canal was the forerunner of canal networks, opened on 17 July 1761, the first in Britain to be built which did not follow an existing watercourse and one which marked the beginning of the canal-building era which ran from 1760 to around 1830.

Francis was an internal cylinder 0-6-0 saddle tank (3068/1917) built by Kerr, Stuart at their Stoke-on-Trent works. Founded in 1881 by James Kerr as James Kerr & Company, later Kerr, Stuart & Company from 1883 when John Stuart joined as a partner. The business started up in Glasgow acting as agents ordering locomotives from established manufacturers, among them Hartley, Arnoux & Fanning. They bought HA&F in 1892, moving to the California Works in Stoke to begin building locomotives. In 1930, on closure, the company's designs, spares, etc., were purchased by the Hunslet Engine Company, Leeds.

Francis had a coupled wheelbase of 12ft 0in, 48in driving wheels and cylinders of 17in x 24in, a precisely stated tractive effort of 19,652 lbs was specified at a pressure of 160 lbs, working weight being 50 tons. Its boiler heating surface amounted to 1,145 square feet, the grate area was 20 ½ square feet. The 217 boiler tubes were 1 ¾in diameter. Injectors were by Gresham & Craven of Salford and Walkden. The factor of adhesion was on the favourable side of optimum, being 5.7. Water capacity was 900 gallons.

Of the Victory-class, the loco served time during the First World War on the Inland Waterways and Docks organisation numbered 12. It was later moved to the Railway Operating Division of the Royal Engineers. They had requisitioned over 600 locomotives of various types from thirteen United Kingdom railway companies including *Francis*, which they re-numbered 603. In 1919, the loco was acquired by the Bridgewater Collieries, being overhauled by them at Walkden in 1921 (spelling below as per original):

Jan 19th	6 Bearing Springs	£120
May 25th	1 Pair Copper Steam Pipes	£52
In 1922 further work was carried out; March 17th	108 Cast Iron Firebars	£25 4*s* 7*d*
May 20th	12 Special Bronze Steps	£47 9*s* 4*d*

In 1923 more extensive work was carried out;

Oct 18th	1 New Copper Firebox fitted in Boiler. Copper tubes renewed and boiler repaired	£899 [£40,194.29 in 2013]

Francis next appears in 1926 after the fitting of six new tyres;

Oct 7th	6 Rolled Steel Tyres	£40 16*s* 2*d*
1927		
Nov 17th	2 sets Metallic Packings	£45 12*s* 8*d*

The loco joined the Manchester Collieries fleet after its formation in March 1929. After a spell at Sandhole Colliery, by the late 1930s, *Francis* was to be found at Mosley Common loco sheds, later (*c.* 1945) moving to Astley Green Colliery.

The NCB loco survey of 1947 stated that *Francis* was in 'Fair condition. Suitable for duty.' It entered Walkden Yard for repairs in late 1947 to early 1948 and again from April 1950 until January 1951 (a new boiler from the Hunslet Engine Co. was fitted). Further visits to Walkden came in 1954, late 1956 and early 1957.

Francis had to move with the times after the Clean Air Act of 1956. Dr Giesl's (of Giesl Ejector fame) representative in this country was a Colonel Cantwell, a visitor to Walkden Yard a number of times. According to Joe Cunliffe, former Walkden Yard Manager (d. 1993), he advised on the application of his smoke reducing ejector design to *Francis*, achieved using materials which happened to be around in the workshops. Joe also recalls that *Francis* was a regular Sandhole Colliery engine before it moved to Mosley Common. He remarked on the 'cow's horn' steam regulator, very handily accessible from both sides of the footplate.

A Giesl ejector and unsightly chimney was fitted to the loco to cut down smoke density around 1964 (see photo). Further repair visits to Walkden took place in 1965 and 1967. *Francis*'s long and varied working life of fifty years came to an end at Astley Green Colliery around late 1967. It was scrapped at Walkden Yard in October 1968.

Fred

Fred at Bickershaw Colliery, Leigh, *c.* early 1960s. Robert Stephenson & Hawthorns (7289/1945). Austerity design, released as WD 71480, 24 April 1946.

Fred seen at Walkden Yard for repairs, 20 October 1963, the Yard's steam crane behind. (P. Eckersley, Brian Warmby)

Fred alongside the towering Bickershaw Colliery waste tips (removed/landscaped 2011–13). Coupled wheelbase 11ft 0in, 51in driving wheels, cylinders 18in x 22in, tractive effort 21,060 lbs at 170 lbs, working weight 48 tons. Note the fibreglass 'flower pot' chimney surround cowl associated with the Hunslet mechanical stoker conversion.

Fred possibly 'laid aside' awaiting its future on the line alongside the loco shed at Walkden Yard *c.* 1966–67. The future was to be a bright one, the loco heading off to the Keighley & Worth Valley Railway in July 1968 and today (2013) at Tyseley Depot.

Fred in Walkden shed, coupling rod removed, after purchase by the Keighley & Worth Valley Railway (July 1968). Note the variable aperture vent plate high up on the smokebox door. Along with the Kylpor exhaust system, this varied the firebox draught dependent on the fuel in use. Little used in practice. (Roger Fielding)

Named after Fred Hilton, for many years Master of the Yard, later renamed Manager of Walkden Yard, in post from 1924–50.

This internal cylinder 0-6-0 saddle tank (7289/1945) was built to the standard Austerity design by Robert Stephenson & Hawthorns Ltd, Newcastle-upon-Tyne. The company was formed in September 1937 when Robert Stephenson & Company, which was based in Darlington, took over the locomotive building department of Hawthorn Leslie & Company, based in Newcastle upon Tyne. RSH became part of English Electric in 1955. Locomotive building at the Newcastle upon Tyne works ended in 1961 and at Darlington in 1964.

The loco was ex-War Department 71480 on 24 April 1946. With a coupled wheelbase of 11ft 0in, 51in driving wheels and cylinders of 18in x 22in, a tractive effort of 21,060 lbs was specified at a pressure of 170 lbs, working weight being 48 tons.

The NCB loco survey of 1947 stated that the loco was working on the Central Railways based around Walkden with some shunting duties, but chiefly on long distance traffic between collieries, washeries and main line junctions. Additional comments were: 'In good condition. These Austerities are doing excellent work.' *Fred* was transferred to Bickershaw Colliery, Leigh, on 7 December 1948. It is known to have returned to Walkden Yard for repairs in mid-1949. It was sent from Kirkless Workshops, Wigan, to Walkden Yard in November 1961, returning to Bickershaw Colliery in May 1962, returning to Walkden again in 1963 and from approximately August 1966 until October 1966. *Fred* was taken out of service from Bickershaw Colliery in about July 1967, returning to Walkden Yard. It was sold to the Keighley & Worth Valley Railway in July 1968, leaving Walkden on 17 December of that year. Currently (2013) at Tyseley Locomotive Works and Steam Depot in the West Midlands.

A double WD delight at Keighley Station. *Fred* on the Keighley & Worth Valley Railway, 26 September 1976. 1931 is a WD 2-8-0 Austerity despatched to the Continent during the war, ending up in Sweden, repatriated 1973. Shortly after this photograph was taken it was withdrawn. After thirty years it was rebuilt as 90733. (Pete Hackney)

Gordon

February 1947, Austerity *Gordon* (Robert Stephenson & Hawthorns Ltd of Newcastle-upon-Tyne, 7288/1945, WD 71479) and *Crawford* (see entry, 1883 Wigan Coal & Iron) pass Howe Bridge pit village, Atherton, heading west towards Gibfield Colliery, along the Eccles–Tyldesley–Wigan line. (Albert Leather)

Named after Gordon Nicholls, colliery manager and Mining Agent (1936 to 1946) for the Western District of Manchester Collieries Ltd.

This ex-WD (71479) Austerity design loco was built by Robert Stephenson & Hawthorns Ltd of Newcastle-upon-Tyne (7288/1945) and delivered to the Longmoor Military Railway for storage in July 1945. It was sold to Manchester Collieries in March 1946, arriving on 8 April. Standard Austerity specifications were a coupled wheelbase of 11ft 0in, 51in driving wheels and cylinders of 18in x 26in, a tractive effort of 21,060 lbs at 170 lbs pressure, working weight 48 tons.

Gordon was sent to Chanters Colliery, Atherton, on 18 April 1946 with its new identification painted on rather than plated. With the arrival of nationalisation in January 1947 the NCB loco survey stated that *Gordon* was currently shunting at Chanters Colliery, Atherton, in good condition and suitable for duty. It was sent to Gin Pit Workshops for repair in 1948. The loco remained at Chanters Colliery until 1954, then underwent repairs at Parsonage Colliery, Leigh, later that year until April 1955. It worked on colliery railways in the Standish, Wigan, area until heading off to Kirkless Workshops, Wigan, for repairs in 1958.

Work followed at Wigan Junction Colliery and Maypole Colliery near Abram, based at Low Hall Colliery loco shed, Platt Bridge.

Gordon returned to Chanters Colliery in summer 1959, then Gibfield Colliery, Atherton, from September 1959 until January 1960, moving once more in September 1960 to Bickershaw Colliery. It was taken out of service and scrapped at Bickershaw Colliery by Jose Holt Gordon of Chequerbent around October 1968.

Grand Duchess

The exotically named early Bridgewater Collieries loco *Grand Duchess* (Manning Wardle, Leeds 484/1874) seen at Sandersons Sidings, west of Greenleach Lane, Worsley, north of Worsley Yard. Thought to have been based at Walkden Yard after opening in 1900. Sold probably for scrap in 1924.

Named after Grand Duchess Marie of Russia (17 October 1853–24 October 1920), Duchess of Edinburgh, Duchess of Saxe-Coburg-Gotha. Married in 1874 to Prince Alfred, Duke of Edinburgh, the year of manufacture of this loco.

An internal 15in x 22in cylinder, 4ft diameter driving wheels, 0-6-0 saddle tank loco built (484/1874) by Manning Wardle & Co. Ltd at the Boyne Engine Works (established 1840) in Jack Lane, Hunslet, Leeds. Little is documented in relation to the working life of the loco apart from the fact that *Grand Duchess* was probably based at Walkden Yard after it opened in 1900.

The Bridgewater Collieries Heavy Expenditure books at the Lancashire Record Office tell us that Class O *Grand Duchess* received a new boiler in 1912, costing £592 (or £58,848 in 2013). The next visit for repairs for which records have survived is 1923, when the loco received '14 Copper Rods and Plates £60 2*s* 10*d*', presumably firebox related. The loco was sold, it is thought, for scrapping in 1924.

Gwyneth

Austerity *Gwyneth* (Robert Stephenson & Hawthorns 7135/1944, WD 75185) at Gresford Colliery (Wrexham) shed, where it was based from 1966 to 1973. Parts of *Gwyneth* survive today in the NRM working replica of the Daniel Gooch 1847 GWR 7ft broad gauge 4-2-2 *Iron Duke* built in 1985.

Gwyneth was built by Robert Stephenson & Hawthorns (7135/1944, WD 75185) to the standard Austerity design and was to have both a colourful life and afterlife. It was delivered to WD Liphook, Hampshire, in May 1944, a base for Canadian troops. It then moved to Burton Dassett base, Warwickshire, in early 1945. In service in France as WD 75185, by May 1945 it ended up in store at Calais, returning to the UK in May 1947.

It was purchased by the NCB and was to be found at NCB Llay Main Colliery, Denbighshire, by approximately July 1947, appropriately named *Gwyneth*. Moved to NCB Gresford Colliery near Wrexham on 12 April 1966. Out of use/spare by *c.* June 1973, it was sent to Walkden Yard in April 1974.

It arrived at NCB Bickershaw Colliery, Leigh, in January 1975. Along with *Respite* it had been allocated to the colliery as replacements for *Spitfire* and *Hurricane*, later withdrawn due to firebox problems. In March 1981 the loco was given to the National Railway Museum as a potential source of parts for a replica loco. It headed off to Resco Railways Ltd, Erith, Kent, where in 1985 the boiler, cylinders and motion work of *Gwyneth* were used to build the working replica of the Daniel Gooch 1847 GWR 7ft broad gauge 4-2-2 *Iron Duke* for the National Railway Museum, York's GWR 150 celebrations.

Harry

Austerity *Harry*, Hudswell Clarke, Leeds (1776/1944, ex WD 71499), is the star of this loco shop photoshoot at Walkden Yard *c.* late 1965. Newly repainted and complete with six firebox secondary air tubes aside the smokebox door to help reduce black smoke. Manager Joe Cunliffe is the prankster third left at the back.

Harry at work at Astley Green Colliery, 9 May 1969. After the late 1965 overhaul the loco was given the name *Harry* after Walkden Yard transport foreman Harry Tweedy. Note the steam ring pipe entering the chimney top, doing a debatable job in this case of thinning the smoke! (Steve Leyland)

4 September 1969. *Harry* crosses the Bridgewater Canal (Leigh Branch) at Whitehead Hall Bridge, east of and close to Astley Green Colliery. The loco is returning from the south tipping area and Liverpool–Manchester line exchange, Astley Green Sidings. Smoke reduction apparently very efficient. (Philip Hindley)

Named after Harry Tweedy, formerly Traffic Chargeman at Sandhole (Bridgewater) Colliery, later Foreman, Central Group Railways, Walkden Yard.

This internal cylinder 0-6-0 saddle tank was built to the standard Austerity design by Hudswell Clarke (1776/1944, ex-WD 71499). The company was founded as Hudswell & Clarke in 1860 in Jack Lane, Hunslet, Leeds.

WD 71499 left the works on 30 November 1944. Its value if being sold at that time was £5,286 (£189,450.24 in 2013). In this case it was delivered new (on loan) to the Ministry of Fuel & Power Darton Opencast site, South Yorkshire. It was then on short term loan to the Garswood Coal Co. Ltd, south Wigan, in 1945 then returned to the Wentworth Opencast Coal Disposal Point, South Yorkshire.

In March 1948 it was dispatched to the Peel Hall Opencast Coal Disposal Point, Little Hulton. Many small opencast sites had operated close-by during the Second World War and into the 1950s. The loco returned to Yorkshire for repairs at Green & Palmers, Barnoldswick, from April 1952 until *c.* September 1952. It then returned to Peel Hall screens. Into Walkden Yard for repairs in late 1955, then late 1956 to early 1957 and early to mid-1962.

The NCB Peel Hall Disposal Point closed in mid-1964, the loco then being transferred to Gin Pit Workshops, Tyldesley, and placed in store.

In late 1965 to early 1966 it headed off to Walkden Yard for overhaul and smoke reduction work in the form of a chimney steam jet (see Philip Hindley's colour photo at Astley Green Colliery of 1969), which it was hoped would decrease the smoke density. The loco emerged in the new 'maroon with yellow lining' colour scheme and the name *Harry*. It worked on the Central Railways and Astley Green Colliery (closed April 1970) until approximately late 1973. It was then taken out of service and laid up at Walkden Yard. Sold on 14 December 1976 to Jose K. Holt & Gordon, scrap merchants, Chequerbent, north of Atherton. They had scrapped several ex-Walkden locomotives over the years but for some reason decided to keep the loco, giving it a coat of red lead, a familiar sight to hundreds of people calling there for a 'weigh in'. Holt Gordon sold it in early 1992 to the Shropshire Collection, owned by John Lees, a collection of over fifty industrial steam, diesel and electric locos based at a garden centre in Shrewsbury. The whole of that collection was sold to individuals in 2001.

Bryn Engineering, Horwich, purchased *Harry* from the collection in 2001. The company was set up by John Marrow Snr, an experienced engineer, apprenticed at the NCB Kirkless Workshops, Wigan, in the locomotive shops. He transferred to the loco shops at Walkden Workshops in 1964, later leaving the NCB to work at the privately owned Quaker House Colliery at Orrell near Wigan (owned by Donald Anderson, see *Lindsay*). While there, he led the restoration of ex-Wigan Iron & Coal Company loco *Lindsay* to working order. His son John, also a trained engineer works with him, the company carrying out specialist engineering work on many aspects of heritage railways stock. 2013 comments on *Harry*'s progress from Bryn Engineering:

Recently we removed the tank and cab from the engine and took out the old tubes. We were amazed to find the boiler in much better condition than was feared. We thought the boiler could have had water in it as the dome cover was left off. However, the barrel is fine with one rivet head missing. The crown stays will require replacement as they are in very poor condition and a new tube-plate will be made. Due to 33 years corrosion through outside storage new side patches on the outer firebox are required. The inner firebox is very good. We are now progressing the boiler repairs. Harry *has the tank, wheels and motion from* Allen, *another Astley Green Colliery locomotive. Mechanically the engine is extremely worn with a very thorough overhaul being required.*

Humphrey

Austerity *Humphrey*, Robert Stephenson & Hawthorns (7293/1945, WD 71484), stands out of use at Walkden Yard on 10 July 1973, loco shed and extension left and right behind. In the foreground are components of coalface hydraulic roof supports, to be refurbished at Walkden. (Cliff Shepherd)

Humphrey stood on death row at Walkden Yard for over seven years, here on 8 April 1970 minus coupling rod and smokebox door. Note the six secondary air inlets aside the smokebox door flange. Behind an Austerity boiler stands on a flat wagon. (Philip Hindley)

Named after Edward Humphrey Browne, mining engineer (1911–1987), Manager, Agent then District Mines Manager by 1940 with Manchester Collieries, Group Production Director for the NW Region for the Ministry of Fuel and Power 1939–1945, later North-Western Divisional Board Production Director, 1946–48. By 1950, he had risen to the heights of Director General of Production, the National Coal Board. Chairman West Midlands Division NCB. President of The Institution of Mining Engineers, 1957–58. Later Sir Humphrey Browne MA CBE MIMinEng, Deputy Chairman of the National Coal Board 1960–1967.

This standard Austerity design internal cylinder 0-6-0 saddle tank (7293/1945) was built by Robert Stephenson & Hawthorns as WD 71484. It was initially sent to the Longmoor Military Railway, Hampshire, for storage. It was sold by the War Department to the NCB on 14 March 1946, arriving at Walkden Yard on 17 April 1946.

The 1947 NCB loco survey stated that the loco was in use at Astley Green Colliery, shunting with some main line working. It also stated the loco was in good condition and suitable for duty. Repairs were required soon after, the loco heading to Walkden Yard for a week in September 1947. More repairs were needed in late 1948.

Humphrey was transferred to Chanters Colliery, Atherton, on 31 December 1948. Around 1949 the loco was working at Astley Green Colliery then moved to Howe Bridge Colliery, Atherton. By April 1950 it could be found at Gibfield Colliery, Atherton, and in 1951 at Howe Bridge Colliery.

Repairs were carried out on the loco at Parsonage Colliery, Leigh, in 1954. Two years later more extensive repairs were carried out at Kirkless Workshops, Wigan, for approximately six months. Into Walkden Yard for repairs in November 1961 for five months, returning to Chanters Colliery. Into Walkden for repairs in May 1965 until November 1965, again returning to Chanters Colliery.

Transferred to Gin Pit, Tyldesley, in July 1966 after the closure of Chanters Colliery, *Humphrey* was moved to Astley Green Colliery around late 1967 until the middle of 1968, although little used. By December 1968 the loco had been laid up at Walkden Yard, but it was to be February 1976 before it was sold to Jose Holt Gordon of Chequerbent for scrapping.

Hurricane

Austerity *Hurricane*, Hunslet Engine Co., Leeds (3830/1955), stands gleaming in the sun at Walkden Yard *c.* June 1964, the loco shed far right. It had been 'upgraded' with the Hunslet Engine Co. gas producer system. Note the smokebox air vent, fibreglass chimney surround and three large firebox secondary air holes.

Above left: Cab plate on *Hurricane* shows it was registered by the Railway Executive as No. 244 in 1952. The cab crest of the NCB North Western Area 'officially' applied to the period April 1967–March 1974, yet occasionally was seen after then beneath the grime.

Above right: Hurricane receives a final lick of paint at Walkden in June 1973, now with the cab crest of the National Coal Board, North Western Area (April 1967–March 1974). Into Walkden from June 1973 to March 1974, the gas producer smoke reduction system of 1964 was removed, its chimney reverting back to the original design.

This Austerity design internal cylinder 0-6-0 saddle tank (3830/1955) was built by the Hunslet Engine Co. Ltd, Jack Lane, Hunslet, Leeds (established 1864). It was delivered to NCB Parsonage Colliery, Leigh, on 13 September 1955, resplendent in green with yellow lining. Transferred to Bickershaw Colliery, Leigh, after July 1959, returning to Parsonage Colliery on 11 November 1959. Into Walkden Yard for repairs on 19 November 1963, being fitted with the Hunslet Engine Co. gas producer system, returning to Bickershaw Colliery on 23 June 1964. In late 1964 the loco was to be found once more at Parsonage Colliery for a short while, returning to Bickershaw on 25 March 1965. Back to Walkden Yard for repair in December 1966, returning to Bickershaw in June 1967.

Further repairs at Walkden were carried out in June 1973, the loco only returning in March 1974 after the gas producer smoke reduction system had been removed, its chimney reverting back to the original design by December 1972. Further repairs were carried out on the loco at Walkden Yard in June 1973. Transferred to Bickershaw Colliery, Leigh, on 6 February 1974, being re-numbered 63 000 407. Once more returning to Walkden Yard by August 1977, it ended up being scrapped there around February 1981. Parts from the loco headed off to the Lakeside & Haverthwaite Railway, where similar Hunslet Austerity loco *Repulse* (3698/1950) now resided. *Repulse* had ended its NCB days at Ladysmith Colliery washery, Whitehaven, in September 1976 and can be seen running today (2013).

James

Named after Alderman James Webb, Mayor of Salford 1939–40 and the first Chairman of the National Coal Board, North Western Division from 1 January 1947.

This Austerity design loco was built (7175/1944, WD 71521) by Robert Stephenson & Hawthorns Ltd, Newcastle-upon-Tyne (est. 1937). Standard Austerity specifications included a coupled wheelbase of 11ft 0in, 51in driving wheels and cylinders of 18in x 26in, a precisely stated tractive effort of 21,060 lbs and a working weight of 48 tons.

It was delivered new to the Longmoor Military Railway, Hampshire, for storage in December 1944. Despatched to France in May 1945 and put into store at Calais until returned to the UK in May 1947. It was purchased by the National Coal Board in June 1947, arriving at Walkden Yard on the 16th along with the identical loco *W.H.R.* (7174/1944, WD 71520), and earmarked for work at Astley Green Colliery. The NCB loco survey of 1947 stated that the loco was being reconditioned and painted.

Usually based at Walkden Yard shed until 1954. Working at Astley Green Colliery from August 1947 to early 1948, known to have been working there also from June to December 1949, and in mid-1950. Into Walkden Yard for repairs on 6 September 1952, on 16 January 1953 for a new boiler, further repairs in June 1954, May 1956 (receiving the boiler off *Respite* HE 3696/1950). Working at Brackley Colliery, Little Hulton, from 1957 until 1962, as a replacement for *Joseph* (HE 0-6-0 T 1456/1924) working dirt to the ever-growing Cutacre Tip (closed February 1968, levelled and landscaped 2013). *James* left Brackley after the colliery closed (22 May 1964), moving to Astley Green Colliery from October 1964. Into Walkden for repairs again in mid-1961, mid-1964 and late 1966. *James* was finally laid up at Walkden *c.* August 1967, being scrapped there in November 1968.

Jessie

0-4-0T *Jessie* stood out of use at Walkden Yard for over five years, here seen on 20 September 1964. Built by the Hunslet Engine Co., Leeds (1557/1927), new to Tyldesley Coal Co.'s Cleworth Hall Colliery. (Charlie Verrall)

Jessie out of use at Walkden Yard in 1966, minus dome cover and buffers. The loco had a coupled wheelbase of 6ft 6in, 45in driving wheels, cylinders of 16ft x 24in, a working boiler pressure of 160 lbs and a tractive effort of 16,400 lbs. Its loaded weight totalled 34 ½ tons. Sold for scrap in February 1967. (Roger Fielding)

This 0-4-0 outside cylinder saddle tank was built by the Hunslet Engine Co., Jack Lane, Hunslet, Leeds (1557/1927). *Jessie* had a coupled wheelbase of 6ft 6in, 45in driving wheels, cylinders of 16in x 24in, a working boiler pressure of 160 lbs and a tractive effort of 16,400 lbs. Its loaded weight totalled 34 ½ tons.

Supplied new to Tyldesley Coal Co. in 1927, their last locomotive to be ordered, *Jessie* replaced a loco by Walker Brothers of Wigan of the same name and went into use at Cleworth Hall Colliery, east of Tyldesley.

After nationalisation the NCB loco survey of 1947 stated that *Jessie* was carrying out shunting duties between Cleworth Hall and the main line. It also stated: 'Firebox & tube repairs needed. High axle loading. Not very suitable.' It was overhauled at nearby Gin Pit Workshops in 1951, in later years at Walkden Yard in 1955 and late 1958. The loco was withdrawn from service in March 1961 and sent to Walkden Yard in September 1961. *Jessie* stood unused for over five years in Walkden Yard, allowing rail enthusiasts ample opportunity to document the effects of nature. It was sold to Maudland Metals Ltd of Preston for scrap in February 1967.

Joseph

Joseph, Hunslet Engine Co. Ltd, Leeds (1456/1924). Probably seen here at Brackley Colliery, Little Hulton, in the 1930s where most of its time was spent. Not your typical industrial locomotive with Walschaerts valve gear and dome mounted 'pop' safety valves. Out of use by late 1958, totally dismantled only by 1968.

Named after Joseph Ramsden (1876–1946), director Wm Ramsden & Sons, Shakerley Collieries Ltd, later director and then chairman of Bridgewater Estates (est. 1923), later director then chairman of Manchester Collieries Ltd (est. 1929) and chairman of the Manchester Racecourse Co.

This outside cylinder 0-6-0 side tank (1456/1924) was built by the Hunslet Engine Co. Ltd, Jack Lane, Hunslet, Leeds (est. 1864), in 1924 The hive of activity which was Jack Lane was also the home from 1860 of locomotive manufacturer Hudswell Clarke. *Joseph* was ordered in late 1923, leaving the workshops on 18 June 1924 painted olive green, lined in yellow and vermillion. The identical loco ordered very soon after *Joseph* was *Bridgewater*, costing £3,160 (£150,289.60 in 2013). *Joseph* being finished earlier cost Bridgewater Collieries slightly less at £3,060.

Both, unusually for industrial locomotives, used Walschaerts valve gear, allowing the driver to fine tune the steam engine's operation in a continuous range of settings from maximum economy to maximum power, a bonus on the punishing Bridgewater system.

Joseph had a coupled wheelbase of 13ft 0in, 45in driving wheels and cylinders of 18in x 22in. Tractive effort of 20,700 lbs was specified at 160 lbs boiler pressure. A maximum of 23,500 lbs was also specified at the insured maximum boiler pressure of 180 lbs, the working loaded weight being 50 tons 3cwt. Overall width was 8ft 8in, maximum axle load 17 tons 8cwt, tube heating surface 891 sq. ft, firebox heating surface 100 sq. ft, grate area 17.87 sq. ft, tank capacity 1,100 gallons, bunker capacity 2 tons.

It wasn't too long before substantial repairs were needed on *Joseph*, a cylinder requiring welding in February 1926, further cylinder repairs being needed on 19 March 1926.

		1926				
Feby	26	Welding Cylinder	Joseph	28	0	0
Mar	18	12 Tyres	"Sultan" & "Alert"	84	6	2
"	19	Repairs to Cylinders	"Joseph"	22	16	0
"	27	1 Saddle Water Tank	"Violet"	160	0	0
May	1.31	35 Copper Rods	"Katharine"	60	5	5
July	30	1 Copper Firebox	do	213	7	0
Octr.	4	6 Rolled Steel Tyres	"Francis"	40	16	2

Joseph in Walkden Yard for cylinder repairs in February 1926, only two years after arriving. Note the variety of work carried out at Walkden on other locos. (Lancs Record Office NCB Bw 20/15)

These repairs didn't hold up for long as the records show a new RH cylinder being ordered on 13 April 1927 for £95 (£4,500 in 2013). It appears the design of the loco was subject to twisting and rocking stresses in the frame when working the Central Railways systems, in turn creating stress on the cylinders.

Remarks by the NCB in their February 1947 survey of locomotives in No. 1 Manchester Area were: 'In good condition. New firebox in stock.' It also stated: 'On Central Railways, Walkden. Some shunting, but chiefly on long distance traffic between collieries, washeries and main line junctions.' Repairs were carried out at Walkden from February to November 1948, when the side tanks were replaced. Other repairs were carried out in mid-1952 to early 1953 and mid-1956 to early 1957.

Joseph was for 90 per cent of its time based at Brackley Colliery alongside *Bridgewater*, with occasional duties at Worsley, and a spell working at Astley Green Colliery is noted by observers from 1955 to late 1958. It went into Walkden Yard in late 1958 for overhaul but the work was stopped in May 1959. It was parked outside the loco shed at Walkden minus boiler, tanks and cab for years, the final recognisable pieces only disappearing in 1968.

Katherine

Katherine, Manning Wardle & Co. Ltd, Leeds (1853/1914), awaits repair at Walkden Yard in the 1930s. Note the balance weight positioning in the wheels in relation to the coupling rod attachment point, the second set of wheels being those linked to the cylinders. (Glen Atkinson)

Katherine at Walkden shed in the 1930s, a glimpse of the Manchester Collieries wagon shop behind. The loco originally had larger side tanks, replaced after an accident heading south from Walkden to Mosley Common Colliery in 1936 when it overturned. Scrapped October 1945.

Named after Lady Katherine Louisa Phipps, wife of Francis Egerton, 3rd Earl of Ellesmere (1847–1914). Katherine died in 1926.

The Bridgewater Collieries Heavy Expenditure books at the Lancashire Record Office tell us that *Katherine* was purchased by Bridgewater Collieries in 1914 for £2,490 (£226,017 in 2013). An internal cylinder 0-8-0 tank loco (1853/1914), it was built by Manning Wardle & Co. Ltd at their Boyne Engine Works (est. 1840) in Jack Lane, Hunslet, Leeds. With 50in driving wheels and inside cylinders of 18in x 26in, it operated at a working boiler pressure of 160 lbs, with a Belpaire firebox in use. It has been stated by railway historians that Bridgewater Collieries engineer Mr Greenhalgh requested modifications which made this loco unique.

Katherine fairly regularly headed into Walkden Yard for repairs. In January 1921 the loco received two cast-iron press blocks; in June 1922, 279 lbs of copper plate was expended on refurbishments to the loco. In October 1923 the records show it received eight new steel tyres costing £71 9*s* 1*d*. In January 1924, a new smoke box door was fitted; in September four spring buffers were fitted. December 1924 saw two new brass injectors fitted. In May 1927, it was time for re-tubing, 145 copper tubes costing £152 5*s* 0*d*. No further major expense is recorded until 1933, where the particular series of records ends (post-Manchester Collieries formation in March 1929).

The loco originally had larger side tanks, replaced it is thought after an accident heading south from Walkden to Mosley Common Colliery in 1936 when it overturned. Apart from temporarily working at Gin Pit, Tyldesley, in 1935 it spent most of its working life on the Central Railways, alternating between Walkden and Sandhole Colliery loco sheds. *Katherine* apparently was plagued with minor defects and turned out to be slow on the long hauls with a rolling motion, which some drivers stated might rock you to sleep! It was scrapped in October 1945.

Kearsley

Kearsley awaits the end at Walkden on 20 September 1964. Of the 'Moss Bay' class by Kerr Stuart (3123/1918) at their Stoke-on-Trent works. At Newtown Colliery north of Swinton, later Ashton Moss Colliery, east Manchester, until closure in September 1959. Scrapped by early 1968. (Charlie Verrall)

This compact outside cylinder 0-4-0 saddle tank of the 'Moss Bay' class was built by Kerr Stuart (3123/1918) at their Stoke-on-Trent works (est. 1881). With a coupled wheelbase of 5ft 6in, 38in driving wheels and cylinders of 15in x 20in, a tractive effort of 15,200 lbs was specified at 160 lbs working pressure, working weight being 31 tons.

Originally working with the Royal Engineers Inland Waterways & Docks Department of HM Government. This was formed in 1915, one of their tasks being to operate barges in France and Mesopotamia. By 1916 they were also working barges across the channel from Richborough, east Kent, to the Continent.

Loco *Kearsley* was sold by George Cohen & Sons Ltd, London, to Bridgewater Collieries on 25 January 1926 for £825. After the First World War the company had won a number of large contracts to dispose of surplus military equipment including munitions amounting to 400,000 tons of high explosives and other shells.

The NCB loco survey of 1947 stated that *Kearsley* was shunting at Robin Hood Sidings, which served Newtown Colliery, north of Swinton. It also stated: 'Good engine for duty when in good condition. General overhaul in hand now.'

Into Walkden Yard for repairs on 16 December 1954 until 25 March 1955, leaving Walkden painted royal blue with yellow lining. Sent to Ashton Moss Colliery, east Manchester, in June 1957. The colliery closed in September 1959. Documentation has not survived to state categorically what happened next to the loco, but presumably *Kearsley* was taken out of service and sent to Walkden Yard, being scrapped in late 1967 to early 1968.

Kenneth

Kenneth (NSR 22/1921, LMS No. 2264) crosses Mather Fold Rd, between Newearth Rd and Ellenbrook Brickworks, *c.* 1960. The stone arched structure visible is the Thirlmere Aqueduct valve house. North of here was a 1930s weigh bridge and cabin for coal trains heading north for coal blending at Ashtons Field Colliery.

Kenneth awaits the end at Walkden in September 1964. The loco had trialled a coal and coke mixture in late 1959 to early 1960 to cut down emissions. Taken out of service around October 1961 and scrapped at Walkden Yard in 1964. (Charlie Verrall)

Named after Miles Kenneth Burrows (1888–1979), the youngest son of Miles Burrows of Fletcher Burrows & Co. Ltd, owners of the Atherton Collieries. Miles was the brother of Robert Burrows (1884–1964), former director of Fletcher Burrows & Co. Ltd and later director then chairman of Manchester Collieries Ltd. Robert Burrows was, at the time of the purchase of *Kenneth* in June 1936 from the LMS railway, a director and later the last chairman of the LMS (by late 1937 Robert Burrows had received a knighthood).

The L Class NSR locomotives

John H. Adams of the North Staffordshire Railway designed the first six L class 0-6-2ST locomotives. These were built in 1903 at the Vulcan Foundry, Newton Le Willows, Lancashire. The class was built at the NSR Stoke works from 1913, the loco gaining experience of heavy freight and heavy passenger work with the NSR. One example, No.158, in 1922 ran the Manchester–Stoke run being timed by the legendary Cecil J. Allen. The load was 310 tons (ten coaches), 158 developing 865hp en route. In total thirty-four examples were built, 1923 being the final year of construction. Mechanical lubricators and exhaust steam injectors were to be initially retained in industrial use.

In the mid-1930s a large number of pre-grouping locomotives were being sold or scrapped, by especially the smaller companies. The LMS decided to dispense with their Class L locomotives as a result of the standardisation policy put in place by Stanier in the mid-1930s. Robert Burrows, having a foot in either camp, seized the opportunity for Manchester Collieries to acquire five examples of the six, which were made available.

Kenneth (NSR 22/1921, LMS No. 2264) appears on the surface to the researcher studying archives to have been an exception according to three separate similar ledger entries once held in the managers offices at Walkden Yard. These stated the loco was built by the Midland Railway Company (1844–1922) at Derby. The accuracy of these official entries has been questioned by certain railway historians as *Kenneth* carries a 1921 LMS/ Stoke works plate.

Kenneth was the first example of the class to arrive at Walkden Yard, on 12 June 1936. By 8 October 1937, the remaining four locomotives had arrived, the full complement being;

NSR built locos

King George VI	(NSR 69/1913, LMS No. 2257)
Sir Robert	(NSR 72/1920, LMS No. 2262)
Kenneth	(NSR 22/1921, LMS No. 2264)

LMS built locos

Queen Elizabeth	(NSR 1/1923 LMS, LMS No. 2270)
Princess	(NSR 2/1923 LMS, LMS No. 2271)

Kenneth was an internal cylinder 0-6-2 side tank locomotive. With a coupled wheelbase of 15ft 6in and overall wheelbase of 23ft 0in, the loco had 60in driving wheels and cylinders of 18½in x 26in. The grate area came to 17.8 sq. ft. A tractive effort of 19,450 lbs was specified for a working pressure of 175 lbs, working weight being 64 tons. The coal capacity was 3½ tons.

Repairs were carried out on the loco in October 1933, amounting to a new firebox and 163 new boiler tubes, the loco history summary stating: 'Barrel 10ft 9in x 4ft 9in diameter, new pressure gauge to 280, 175 working pressure.'

The loco had been put to work on the Central Railways with some shunting work, but mainly long distance duties between collieries, washeries and main line junctions. The NCB loco survey of 1947 stated that Kenneth was in good condition. After the Second World War it received a new boiler, rejoining the fleet in October 1945. The loco was in Walkden Yard for repairs in 1951, 1952, 1954, 1955 and 1956. After the Clean Air Act of 1956, Walkden Yard began to address the problem of smoke and the regular complaints from trackside residents. *Kenneth* is known to have trialled a coal and coke mixture in late 1959 to early 1960. The experiment was not to be successful and ended. The loco was taken out of service around October 1961 and was scrapped at Walkden Yard in 1964.

King

Loco *King*, Andrew Barclay Sons & Co. Ltd (1448/1919), at Walkden Yard, behind it sideways on is a colliery winding skip. Based at Ashton Moss Colliery, east Manchester, on nationalisation in 1947, Bank Hall Colliery, Burnley, late 1959, Ladysmith Colliery washery, Cumberland, in 1972 and scrapped there around October 1975.

This outside cylinder 0-4-0 saddle tank (1448/1919) was built by Andrew Barclay Sons & Co. Ltd. They first set up their Caledonia engineering workshop in 1840 in Kilmarnock, Scotland. After a long period of operation the company was acquired by the Hunslet group in 1972 and renamed Hunslet-Barclay.

With a coupled wheelbase of 5ft 6in, 41in driving wheels and cylinders of 14ft x 22ft, a tractive effort of 12,600 lbs was specified at 160 lbs working pressure, working weight being 30 tons. The loco was bought by the Chamber Colliery Co. of Oldham in 1934 and used at their Ashton Moss Colliery, east Manchester.

The 1947 NCB loco survey stated that *King* was working at NCB Ashton Moss Colliery. It also stated that it was suitable for duty when in order and that the engine, frame and boiler were being overhauled (arrived Walkden Yard 28 April 1947). A new boiler complete with copper firebox was delivered from Houghs of Wigan.

In possibly the longest sentence ever written the Walkden Yard record for this overhaul is as follows:

Engine lifted and wheels taken out, tyres renewed, loco carbon tyres were used, crank pins turned up, axle boxes and axle box brasses renewed, axle box keeps renewed, bearing springs, brackets & pillars renewed, side rods machined out, new brasses and cotters fitted, big and little end brasses and cotters renewed, R & LH cylinders bored, new piston heads, rings, piston rods renewed, slide bars renewed and lined up, crosshead liners renewed, crosshead pins renewed, slide valves lined and machined, valve spindles renewed, valve motion ground out and motion pins renewed, keys renewed in eccentric sheaves, eccentric straps closed, brake work renewed, drawbars and links renewed, cylinder drain taps renewed, lubricator to cylinder repaired, frame repaired, new boiler fitted in frame, water gauge taps renewed, boiler mountings overhauled, new safety valves fitted, new dome cover fitted, R & LH injectors overhauled, injector check valve renewed, steam pipes fitted, exhaust pipe renewed, clothing on boiler renewed, tank and cab repaired, eyeglasses renewed, engine painted and tested.

Returned to Ashton Moss September 1947

In September 1959, Ashton Moss Colliery closed, *King* being transferred to NCB Bank Hall Colliery, Burnley. The NCB loco survey 1962 update stated that the loco was still working in the Burnley coalfield. Bank Hall Colliery closed in March 1971, the loco being sent to Ladysmith Colliery washery in the Cumberland coalfield in 1972 after overhaul at Walkden Yard. Photographs by enthusiasts now available on the internet show the diminutive *King* at Ladysmith with a full-size Giesl ejector chimney, a very strange sight. It was scrapped at Ladysmith around October 1976, the washery having closed in 1975.

King George VI

King George VI was built by the North Staffordshire Railway at their Stoke-on-Trent works (69/1913, LMS No. 2257 until May 1937). Seen *c.* late 1950s to early 1960s on chocks in the shed at Walkden. The old LMS No. 2257 numbers can still be seen on the coal tender and smokebox door. (Glen Atkinson)

Above left: A fascinating shot from the late 1950s to early 1960s at Walkden of *King George VI*. Taken before the smokebox secondary air adaptation carried out in 1964. In the foreground its chimney and spark arrestor mesh, middle right two pistons end-on. (Dave Ingham)

Above right: King George VI working at Sandhole Colliery washery, 26 September 1964. Note the three secondary air holes in the smokebox side and steam ring supply to the chimney top. Dismantled at Walkden by August 1965, its bunker and other fittings were used on *Sir Robert*. Scrapped by May 1966. (Philip Hindley)

As mentioned in the description of loco *Kenneth*, Manchester Collieries director Robert Burrows (1884–1964) was at the time of the purchase of *King George VI* and the other NSR/LMS locomotives from the LMS (June 1936–October 1937) a director of the LMS. Robert Burrows had been for a number of years a close friend of the Duke of York, the Duke occasionally staying at his Bonis Hall home in Prestbury, Cheshire. By the time of the loco purchases the Duke was King George VI (acc. 20 January 1936), hence the naming. The loco arrived at Walkden in May 1937. The favour was to be returned, Robert becoming Sir Robert Burrows later in 1937.

King George VI was an internal cylinder 0-6-2 side tank locomotive (69/1913, LMS No. 2257 until May 1937) built by the North Staffordshire Railway at their Stoke-on-Trent works. With a coupled wheelbase of 15ft 6in and overall wheelbase of 23ft 0in, the loco had 60in driving wheels and cylinders of 18½in x 26in. A tractive effort of 19,450 lbs was specified for a working pressure of 175 lbs, working weight being 64 tons.

The loco worked on the Central Railways, based at Walkden Yard shed up to and beyond coal industry nationalisation in January 1947. It is known to have been re-boilered in 1948, the boiler supplied by the Hunslet Engine Co. Ltd, Leeds. In again for repairs in 1951 and 1956. Based at Sandhole Colliery in 1957. Back into Walkden for repairs in 1959, 1961 and late 1962. It was dismantled at Walkden by August 1965, its bunker and other fittings being used on *Sir Robert*, and was scrapped by May 1966.

Geoff Hayes, in his *Collieries in the Manchester Coalfields* of 1986, recalls that the NSR locomotives were excellent performers, their boilers steamed very freely, urged by a draught that shot red hot cinders to an incredible height above the chimney.

Lindsay

Lindsay, built 1887, Wigan Coal & Iron Co. Ltd. Coupled wheelbase 11ft 10 ½in, 51 ½in driving wheels, cylinders 16in x 20in, working boiler pressure of 125 lbs, tractive effort 9,340 lbs, working weight 35 tons. Seen at John Pit, Standish, 20 July 1957. (Jim Peden)

Lindsay ended its working life at Hafod Colliery, North Wales, around March 1968. Here seen at Maudland Metals, Preston, where in 1976 it was to be rescued and restored at Quaker House Colliery, south Wigan. Loaned to Steamtown, Carnforth (closed as a museum in 1997), now the base for West Coast Railways. On site 2013. (Preston Digital Archive)

Named after Alexander Lindsay, 25th Earl of Crawford, 8th Earl of Balcarres (1812–1880), *Lindsay* was the fourth (it is presumed) locomotive to be so-named to work among the Haigh and Aspull collieries, north of Wigan. The first *Lindsay* is thought to have been the 0-4-0 tender loco of *c.* 1842–44 belonging to the Earl, which suffered a boiler tube explosion on 4 February 1863. The second *Lindsay*, a six-coupled tender loco using parts from the first example, had only just been constructed when working on 10 December 1867. It was derailed and the driver killed. It is now presumed by various railway historians that a rebuild of the damaged loco created the third *Lindsay*, the work being carried out before the arrival of the fourth *Lindsay*.

The fourth *Lindsay*, which survives today, is an internal cylinder 0-6-0 saddle tank built in 1887 by Wigan Coal & Iron Co. Ltd at their Kirkless, Aspull, works. With a coupled wheelbase of 11ft 10½in, 51½in driving wheels and cylinders of 16in x 20in, a working boiler pressure of 125 lbs (insured to 150 lbs) produced a tractive effort of 9,340 lbs (11,330 at 150lbs), working weight being 35 tons.

Wigan-based consultant mining engineer Cornelius McLeod Percy visited the LNWR works, Crewe, for WC&ICo. in 1864, obtaining locomotive designs, accompanied by engineering staff from WC&ICo. A standard WC&ICo. design of 0-6-0 with saddle tank and internal 16in x 20in cylinders was arrived at, loco *Lindsay* being an example (see loco *Crawford*).

It is known that loco *Lindsay* was loaned to Bridgewater Collieries during the First World War, although I have not been able to track down any archives recording this. The loco joined the fleet of Wigan Coal Corporation in 1930 (see *Crawford*). By January 1930 it was at work at Parsonage Colliery, Leigh, and once more known to have been working there in November 1930. In 1940 it was once more at Parsonage Colliery and also in 1943. The NCB loco survey of 1947 stated that *Lindsay* was working at Parsonage Colliery, Leigh, and in fair condition. It is known to have gone into Kirkless Workshops for repair on 12 April 1956, this time returning to John Pit, Standish, around August or September 1956. It is known to have worked at Howe Bridge Colliery, Atherton, for a short while until 18 April 1957, then temporarily at Ince Moss Colliery, Wigan, from June until August 1961 (closed June 1962). Into Kirkless Workshops, Wigan, for repair on 12 December 1961, returning to Chisnall Hall Colliery, Coppull, on 14 May 1962. It was moved from Chisnall Hall to the Standish area around January 1963. After the last colliery to work in the Standish area, Robin Hill Drift, closed on 29 November 1963, *Lindsay* was transferred back to Chisnall Hall Colliery, and is known to have been working there between April 1964 and July 1964. Chisnall Hall Colliery closed in March 1967. *Lindsay* then makes quite a move in August 1967, to Hafod Colliery, North Wales (closed 11 March 1968).

Lindsay was never put into use at Hafod and the coupling rods never refitted. It stood in the Maudland Metals scrapyard in Preston for many years before being purchased in December 1976 by Donald Anderson, mine owner, mining historian, and his colleagues. The engine was restored to working order at Donald Anderson's Quaker House Colliery, near Leyland Green, south of Wigan, by the colliery company's fitters, including John Marrow. At the time of writing, John and his son were operating Bryn Engineering at Horwich, undertaking steam locomotive engineering work, on *Harry* for example.

Lindsay was repainted in the vivid red livery used by Wigan Coal & Iron Co. Ltd in the late nineteenth century. It was sent, on loan, to Steamtown Railway Museum at Carnforth, which sadly closed to the public in 1997 and is now the base for West Coast Railways (see colour section). I am told the loco can only now be viewed on site by appointment (2013).

Madge

Madge in ex-works photo 'colours'. Purchased in 1906 by Bridgewater Collieries from Nasmyth, Wilson & Co., Bridgewater Foundry, Patricroft, Manchester (782/1906), for £2,299 (or £220,106 in 2013). Supplied with cast iron brake blocks, replaced with wooden ones by Bridgewater Collieries.

Madge photographed shortly after purchase in 1906. *Madge* was to be based at Mosley Common Colliery after formation of Manchester Collieries in 1929. Scrapped at Gin Pit Workshops, Tyldesley, in 1946, various parts ended up as spares for the identical engine *Violet*, which worked at Chanters Colliery, Atherton, until 1962.

The naming of *Madge* is problematical as no Margarets (or Madges!) were present in the Worsley Ellesmere/Egerton dynasty at the time of purchase. Its sister engine *Violet* was purchased two years later, having a more obvious link with the family (see *Violet*).

The Bridgewater Collieries Heavy Expenditure books at Lancashire Record Office tell us that loco *Madge* was purchased in 1906 by Bridgewater Collieries from Nasmyth, Wilson & Co. (est. 1867), Bridgewater Foundry, Patricroft, Manchester (782/1906), for £2,299 (or £ 220,106 in 2013). This internal cylinder 0-6-0 saddle tank had a total wheelbase of 15ft 8in, 49in driving wheels and cylinders of 17in x 24in, a precisely stated tractive effort of 19,251 lbs was specified at 85 per cent of 160 lbs boiler pressure. The boiler measured 11 ft long x 4.0875 ft diameter, 172 tubes in place of 2in outside diameter. The heating surface amounted to: tubes 1,018 sq. ft, firebox 91 sq. ft, total 1,109 sq. ft. Injectors comprised two of the Gresham and Craven combination design. The blast pipe nozzle was of 4.5in diameter, water capacity 1,200 gallons, coal capacity 1.5 tons. Height to chimney top was 12ft 7in, length over buffers 30ft 6in, width 8ft 3.5in. The locomotive's working weight was 45.5 tons and factor of adhesion 5.29. The loco came equipped with cast iron brake blocks, replaced with wooden ones by Bridgewater Collieries.

The valuation of locomotives by Bridgewater Collieries of 31 December 1920, put the value of *Madge* at £2,000; not too bad a depreciation from £2,299 after fourteen years, albeit with regular maintenance at substantial outlay ensuring the loco kept its value. The records show that *Madge* visited Walkden Yard in February 1923 for two new cast iron cylinders costing £119 5*s* 0*d*. Further outlay was required in February 1924 when the loco received a new saddle tank at £1,600. In May of that year a new boiler with copper firebox was required, costing £1,136 15*s* 6*d*.

In Manchester Collieries days, post-1929, it is known that the loco was based at Mosley Common Colliery shed. In March 1938, *Madge* was transferred to Gin Pit Workshops, Tyldesley, intended for Chanters Colliery, Atherton. The loco returned to Gin Pit from Chanters later that year for repairs. *Madge* was scrapped at Gin Pit Workshops in 1946, various parts ending up as spares for the identical engine *Violet*, which carried on working at Chanters Colliery until 1962.

Outwood

Outwood seen at Walkden on 14 January 1962 after scrapping had begun, the boiler being retained for stationary use. (P. Eckersley, Brian Wharmby)

Outwood (formerly *Outwood No.1*) was one of two so named. The longest survivor was this outside-cylinder 0-6-0 saddle tank built by Peckett, works number 923/1901. Cylinders 14in bore x 20in stroke, coupled 3ft 7in diameter wheels on a 10ft wheelbase. Working pressure 150psi, tractive effort at 75 per cent pressure was 10,250. The boiler had 132 copper tubes, 1 1/2in internal diameter, gauge 1/8in. Tank capacity 800 gallons, loco weight 29 tons.

Clifton & Kersley Coal Co. Ltd took over Outwood Colliery in 1909 from Thomas Fletcher & Sons Ltd, locos *Outwood No.1* and *Outwood No.2* being on the books. Outwood Collieries Ltd was later taken over by Manchester Collieries on formation in 1929 (*Outwood No.2* was an 0-4-0ST by Peckett, 935/1901, disposed of by 1922). Based by Manchester Collieries at Outwood washery, the loco was overhauled at Walkden from 22 October 1934 to 14 February 1935, being returned to Outwood. Maintenance after this date included, apart from annual boiler inspections:

August 1943 New fusible plugs
September 1943 New ash pan
January 1945 Piston rods and rings renewed, slide bars machined and lined up, crosshead slippers renewed. Big and little end brasses renewed. Side rod cotters renewed. Injectors overhauled. Blow off valves fitted on boiler
May 1946 Left hand side rod brasses renewed
June 1946 Axle boxes lined up, brasses bedded in. Six crank pins renewed, side rod brasses renewed. Big and little end brasses renewed, side rod coupling pins and crosshead pins renewed. Top crosshead liners renewed and slide bars lined up. piston rods repacked. Steam brake joints remade. Brake work repaired.
October 1946 (At Outwood shed). Engine lifted, wheels taken out, driving axle renewed.

The NCB survey of locos of 1947 had stated that it was in fair condition and based at Outwood washery. From April to December 1947 it was overhauled and repainted. Overhauled again and repainted, September 1948 to May 1949. In for minor repairs July 1949, November 1949 and in March 1950 overhauled. A new boiler was fitted by 3 December 1953, working pressure 150 lbs, pressure tested to 250 lbs.

The loco was next to be based temporarily at Wheatsheaf Colliery (Pendlebury, north of Swinton) screens from March to June 1954. Sent to Walkden in August 1956 to be scrapped. It was fitted with a chimney extension and used as a stationary boiler by September 1957. Partly scrapped by January 1962 by the Central Wagon Co. Ltd, the boiler was removed for continued stationary use at Walkden. The loco's frames were scrapped in April 1962.

Princess

Princess, seen here at Walkden in the late 1930s, was an internal cylinder 0-6-2 side tank locomotive (?/1923, LMS No. 2271 until October 1937) built by the LMS at the old North Staffordshire Railway works, Stoke-on-Trent. Its old LMS 2271 number is still in place on the smokebox door. (Glen Atkinson)

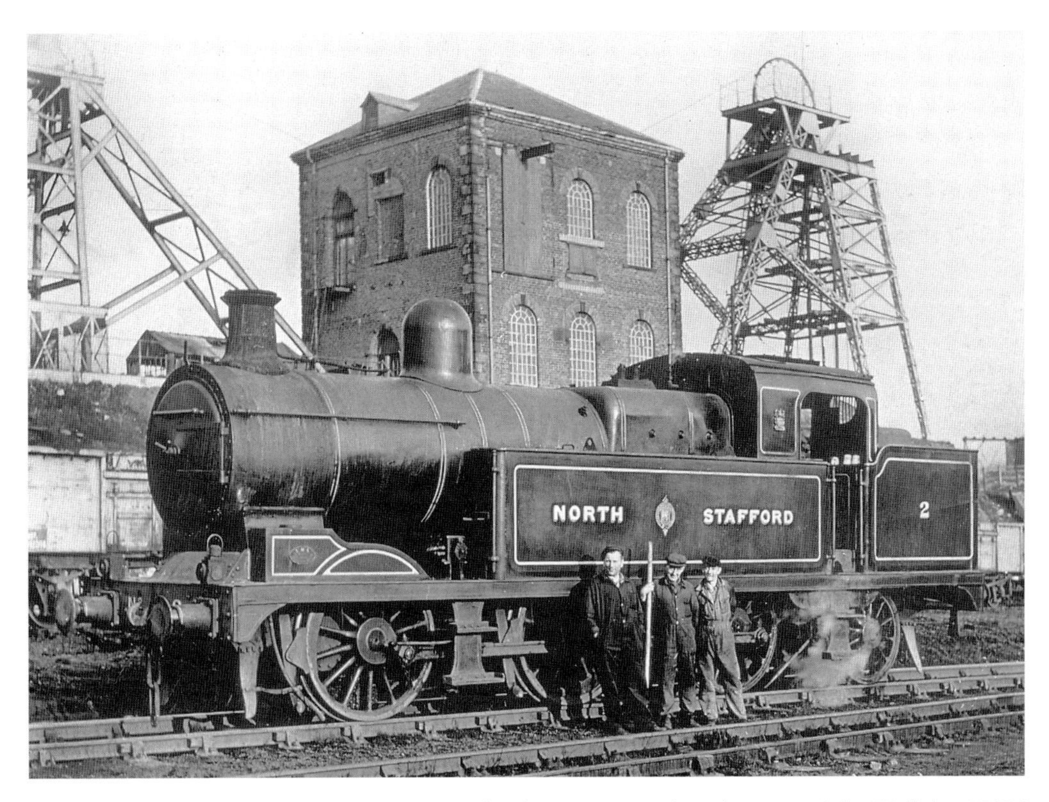

1960 was to be a special year for *Princess*. The decision was taken that one of the Walkden NSR/ LMS locomotives would take part in the City of Stoke-on-Trent Golden Jubilee Celebrations (of the Federation of the Six Towns), March 31 – 30 September 1960. The railway exhibition was held at the London & North Western yard, Stoke-on-Trent, from 11 May until 24 May 1960. *Princess* was chosen (even though built under the LMS in 1923!), the loco being overhauled at Walkden and sent to Crewe on 22 April for a repaint. By late April 1960 it had been repainted in NSR madder lake livery, emerging as North Stafford 2, and appeared at the Stoke-on-Trent exhibition, returning to Walkden in late May 1960.

Here, after return to Walkden, its regular crew pose for the photographer, from the left driver for three years Cyril Almond, Henry Brown brakesman with forty years service and Horace Athern the fireman (ex-BR).

Behind stands the imposing Ellesmere Colliery engine house of 1866, originally housing a Nasmyth, Wilson steam winder of 30in x 50in. No.1 on the left was sunk to 275 yards, No. 2 sunk down to 415 yards. The twin wooden headgears were replaced in the 1950s with the steel examples visible. The colliery had ceased coal production in 1921, the shafts being retained for ventilation and access to the Worsley underground canal system until the closure of Mosley Common Colliery in February 1968. The engine house and headgears were demolished despite attempts to preserve them in November 1968.

Princess and Walkden Yard loco shop men after the return of the locomotive around May 1960.

Walkden Yard shed around May 1960. Second from the left is 47669 (0-6-0TIC Hor/1931), a BR loco brought in to replace *Princess/North Stafford No.2*, which was on display at the Stoke-on-Trent railway exhibition from 11 May until 24 May 1960. (Alan Tyson)

A pristine *Princess/North Stafford No.2* seen at Walkden shed on 31 July 1960 having returned from the Stoke-on-Trent celebrations shortly after 24 May. (P. Eckersley, Brian Wharmby)

Walkden Yard shed, late 1964 to early 1965. In 1964 *Princess* was overhauled receiving the frames, cylinders and wheels from *Sir Robert*. The resulting loco was taken out of service by August 1965. *Kenneth* to the right awaits scrapping. (Roger Fielding)

Princess looking impressive (even more so in colour!) outside Walkden shed on 19 September 1964, by now sporting the frames, cylinders and wheels of *Sir Robert*. Some say the loco should after then have been regarded as *Sir Robert*. (Philip Hindley)

Named after Princess Elizabeth, the future Queen Elizabeth II, daughter of Queen Elizabeth, wife of King George VI. Similar (apart from originally being superheated) to the other North Staffordshire/LMS locomotives earlier described: *Kenneth* and *King George VI*, *Princess* was an internal cylinder 0-6-2 side tank locomotive (?/1923, LMS No. 2271 until October 1937) built by the LMS at the old North Staffordshire Railway works, Stoke-on-Trent.

Worth noting is the fact that *Princess* and a fifth similar loco to be delivered to Manchester Collieries, *Queen Elizabeth*, were actually built in 1923. Agreement had been passed for the NSR to join the LMS on 1 January 1923 but the NSR did not for legal reasons amalgamate until 1 July 1923. Being constructed after 1 January 1923 these locos are not then to be regarded (I am told by respected authorities) as true North Staffordshire Railway locomotives.

With a coupled wheelbase of 15ft 6in and overall wheelbase of 23ft 0in the loco had 60in driving wheels and cylinders of 18½in x 26in. A tractive effort of 19,450 lbs was specified for a working pressure of 175 lbs, working weight being 64 tons. Grate area 17.8 sq. ft, coal capacity 3 ½ tons.

After the Second World War, new boilers were ordered for the NSR/LMS locomotives including *Princess*, dealt with in November 1946, also receiving new side and rear water tanks, a new bunker, new cylinders and slide valves. The NCB loco survey of 1947 justifiably stated that *Princess* was in good condition, its duties on the Central Railways being some shunting with mainly long distance work between collieries, washeries and

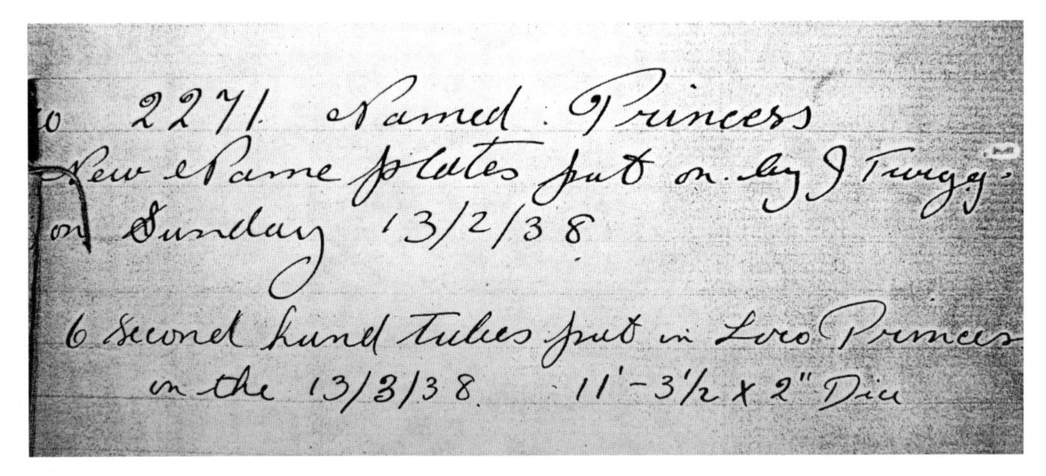
2271 Named: Princess
New Name plates put on by J Twigg
on Sunday 13/2/38

6 Second hand Tubes put in Loco Princess
on the 13/3/38. 11'-3½ x 2" Dia

The Walkden Yard entry for 2271 *Princess* (arrived 8 October 1937). After inspection and test runs its new nameplates were in place by February 1938. Six eleven-foot-long boiler tubes were installed at Walkden a month later.

main line junctions. The years following nationalisation in 1947 relating to *Princess* are not well documented, presumably carrying on its previous duties.

1960 was to be a special year for *Princess*, rather better documented than those previous. The loco is known to have been at Walkden for repairs on 17 January, being stripped down to its frames. By 21 February it was back working at Sandhole Colliery. It was back into Walkden by 28 February, marked up for repairs and noted as still being there in April.

The decision was taken that one of the Walkden NSR/LMS locomotives might take part in the City of Stoke-on-Trent Golden Jubilee Celebrations (of the Federation of the Six Towns). These ran from 31 March until 30 September 1960. The railway exhibition was held at the London & North Western Yard, Stoke-on-Trent, from 11 May until 24 May 1960. *Princess* was chosen (even though built under the LMS in 1923), the loco being overhauled at Walkden and sent to Crewe on 22 April for a repaint. By late April 1960 it had been repainted in NSR madder lake livery, emerging as *North Stafford* 2, and appeared at the Stoke-on-Trent exhibition, returning to Walkden in late May 1960. In its absence British Railways provided Walkden with a six-coupled side tank loco numbered 47669.

In 1964 the loco was overhauled at Walkden Yard, receiving the frames, cylinders and wheels from *Sir Robert*. It has been suggested to me that the resulting locomotive should then have been entitled *Sir Robert*, as it was built up around its frames; the debate I am sure will continue!

Princess was taken out of service by August 1965, heading off to Shugborough Museum, Staffordshire, arriving on 5 June 1967. After a spell at Chatterley Whitfield Mining Museum, Stoke-on-Trent, after 1979 (closed as a working mine 1976, opened as a mining museum 1979, closed 1991) the locomotive headed off *c.* 1994 (see photo in colour section) to the Churnet Valley Railway at Cheddleton, near Leek, Staffordshire. In March 2005, it joined the collections of the National Railway Museum at Shildon, County Durham (opened 2004). Today it is on display in non-running order, complete I am told with the dents, knocks and scratched paintwork dating back to its final NCB days.

Queen Elizabeth

Queen Elizabeth at Walkden in 1943 (arrived 8 October 1937). Built by the LMS at the former North Staffordshire Railway, Stoke-on-Trent works (?/1923, LMS No. 2270). Note colliery headgear pulley wheels stacked behind and loco wheel tyres right foreground. Scrapped mostly by May 1957. (Harry Townley)

Named after Queen Elizabeth, wife of King George VI. Similar to the other North Staffordshire/LMS/Midland Railway Co. locomotives, *Kenneth*, *King George VI*, *Princess* and *Sir Robert*, *Queen Elizabeth* was an internal cylinder 0-6-2 side tank locomotive (?/1923, LMS No. 2270 until October 1937) built by the LMS at the former North Staffordshire Railway, Stoke-on-Trent, works. With a coupled wheelbase of 15ft 6in and overall wheelbase of 23ft 0in the loco had 60in driving wheels and cylinders of 18½in x 26in. A tractive effort of 19,450 lbs was specified for a working pressure of 175 lbs, working weight being 64 tons.

It is known to have entered Walkden Yard for repairs on 6 December 1945 to receive a new boiler (the boiler coming from the Hunslet Engine Company, Leeds), the work on the loco taking until January 1947 to complete.

By 1952, *Queen Elizabeth* required expensive repairs. It was decided to withdraw it by August 1952, its boiler and other parts being earmarked for spares. The loco was being partly dismantled by April 1954: its chassis converted to a boiler carrier, the trailing wheels were removed and the rear end of the frames cut off. The remains of the loco were scrapped in May 1957. The chassis itself was scrapped in May 1963.

Radcliffe

Ex Andrew Knowles & Sons Ltd *Radcliffe*, Manning Wardle & Co. Ltd, Atlas Works, Bristol (1192/1890), probably began work at Allens Green Colliery, Radcliffe. Here seen at Wheatsheaf Colliery, Pendlebury, in February 1952, a few months before scrapping. (Alex Appleton)

This internal cylinder 0-6-0 saddle tank was built by Manning Wardle & Co. Ltd at their Atlas Works in Bristol (1192/1890) and delivered new to Andrew Knowles & Sons Ltd. The company had been established in 1859 by John Manning, C. W. Wardle and A. Campbell at the Boyne Engine Works, Leeds.

With a coupled wheelbase of 10ft 9in, 36in driving wheels and cylinders of 11in x 16in, a tractive effort of 4,030 lbs was specified, working weight being 18 tons. The early working life of the loco appears not to be documented in surviving archives, but can be pinned down to the Knowles' collieries near Radcliffe, namely Withins Lane, Hagside, Allens Green and Coney Green, all within a mile of each other and all active at the time of the purchase of loco *Radcliffe*. Narrowing down the list to collieries sunk closest to the date of purchase of the loco leaves Allens Green and Coney Green. Coney Green had no direct rail connection to the Lancashire & Yorkshire Railway, leaving Allens Green Colliery as favourite.

The Knowles' collieries in Radcliffe closed around 1935, loco *Radcliffe* then moving to the Knowles' collieries near Swinton and Pendlebury, namely Wheatsheaf, Clifton Hall, Agecroft and Pendleton. It appears that *Radcliffe* spent most of its working life after the move at Wheatsheaf Colliery, Pendlebury, with periods also at Clifton Hall Colliery. On nationalisation of the coal industry in January 1947, the NCB loco survey stated that the loco was in very poor condition; it was to be scrapped around August/September 1952.

Renown

One of five Hunslet Austerities ordered by the NCB in 1950 and delivered to Walkden Yard, *Renown* (3697/1950), here seen at Bedford Colliery, east of Leigh, where it worked in the early, mid- and late 1960s. Its very short working life ended with scrapping at Walkden in October 1968.

In 1950, the North Western Division of the National Coal Board ordered five of the highly regarded Austerity class 0-6-0ST internal cylinder shunting locos from the Hunslet Engine Co., Jack Lane, Leeds. Delivered to Walkden Yard, they were (apart from *Respite*) named after past and present battleships: *Rodney* (3695/1950) for Bickershaw Colliery, Leigh, *Respite* (3696/1950) for Astley Green Colliery, *Renown* (3697/1950) for Parsonage Colliery, Leigh, *Repulse* (3698/1950) and *Revenge* (3699/1950)

Renown was delivered to Walkden Yard on 4 November 1950. It was initially sent to Parsonage Colliery, Leigh. In August 1951 it was moved from Parsonage Colliery to Gin Pit, Tyldesley. It was decided to retain the loco for the No. 1 (Manchester Area) fleet on 1 January 1952, mainly as a spare it seems from the working records. It went into Gin Pit Workshops, Tyldesley, for repairs in June 1955. Repairs afterwards would be at Walkden Yard as Gin Pit Workshops phased out loco repair work, the loco going in for repair in 1956.

Steve Oakden recalls in his article for the Industrial Railway Society 'Working on the Gin Pit Railway' (*Industrial Railway Record* No. 114, Sept. 1988) that in the early to mid-1950s *Renown* needed some home grown adjustments to make her steam freely: 'a jimmie was placed in the blast pipe'. This was a restriction, a metal plate or a wire strung across the aperture that had the effect of increasing the air flow. It was an illegal practice due to the danger of parts falling in and possibly damaging the valve gear.

A particular driver, George Robinson, is highlighted as being a good driver, at times coupling up to twenty-two loaded sixteen-tonners (352 tons plus loco 48 tons total 400 tons). He managed the final incline at St Georges Colliery by deftly easing off the regulator between each beat while working the sanding lever with his other hand!

Renown is known to have headed to Walkden Yard for repairs by 29 June 1958. It was back on the Gin Pit railways by 26 September 1958. Working on the Walkden Central Railways by December 1960, back to Gin Pit by May 1961. Based at Bedford Colliery, Leigh, by April 1962, returning once more to Gin Pit by February 1963. Based at Walkden Yard by October 1963 and in for repairs a month later.

It left Walkden for Gin Pit in early 1964 (Gin Pit itself had ceased winding coal in September 1955). From May to November 1965 it was based at Bedford Colliery as *Humphrey* was at Walkden for repairs. During August and September 1965 the loco could be found shunting low-grade coal originally from Chanters Colliery, Atherton, away from Nook Pit washery, Tyldesley. The coal had arrived by road to Nook Pit. The Chanters washery was not working at the time and only had a few more months' life left, closing in June 1966, so was never repaired.

Renown went into Walkden Yard for repairs in August 1966. In March 1967 we find the loco once more working at Bedford Colliery, until the colliery closed in October 1967. It moved to Gin Pit on 21 October 1967, leaving there in November 1967 to be stored at Walkden until August 1968. The loco worked at Astley Green for a short while in September 1968, its very short working life coming to an end at Walkden Yard in October 1968 when it was scrapped.

Repulse

Walkden Yard open day, 14 August 1966. Gleaming Austerity *Repulse* (Hunslet Engine Co., Leeds, 3698/1950), is on view to the public who could now inspect its Giesl conversion. *Wasp*, *Harry*, *Sir Robert* and *North Stafford* 2 were also accessible at close quarters. (Brian Wharmby)

Austerity *Repulse*, Hunslet Engine Co., Leeds (3698/1950), at Walkden shed *c.* 1968–70, the NCB NW Division crest on the cab side. The loco was stored at Walkden in early 1970, overhauled in early 1973, then sent to Ladysmith Colliery washery, Whitehaven, by 28 May 1973.

Walkden Yard line up *c.* late 1968, Bridgewater Offices and clock beyond. From the left: Austerity *Repulse* with Giesl conversion; Hunslet Austerity *Wizard* (3843/1956), also recently converted; and *Sir Robert* (NSR 72/LMS No. 2262). (Roger Fielding).

Repulse's nameplate removal would have proven a challenge for the determined souvenir hunter. Today it can be seen in its rightful place at the Lakeside & Haverthwaite Railway in Cumbria. (Roger Fielding)

Repulse shunts internal wagons at Astley Green Colliery on 27 March 1968, the colliery washery and coal preparation plant behind, the Bridgewater Canal (Leigh Branch) in the foreground. (Philip Hindley)

This 0-6-0ST internal cylinder Austerity design (see specifications in main text earlier) loco was built by the Hunslet Engine Co., Jack Lane, Leeds (3698/1950), one of five identical locos ordered by the NCB North Western Division. It was delivered to Walkden Yard on 16 November 1950.

The loco was based at Walkden Yard until as late as early 1968, with spells working at Brackley Colliery, Little Hulton (1952), Sandhole Colliery (1957, 1958) and Astley Green Colliery (1968, 1969). Repairs are known to have been carried out on *Repulse* at Walkden Yard in mid-1954, early 1957 and late 1960 to early 1961. A Giesl ejector was fitted in October 1962 and the Hunslet Engine Co. underfeed stoker was fitted by May 1967.

The loco was stored at Walkden in early 1970 and overhauled in early 1973, then sent to Ladysmith Colliery washery, Whitehaven, Cumbria, by 28 May 1973. As with a number of former NCB colliery locos, the prospect of being sent to the cliff top system at Ladysmith washery (which washed small coal from Haig Colliery, now a museum) was often a short duration death sentence and might have been the death knell for *Repulse* due to the savage treatment, climate and lack of proper maintenance of locos (I am reliably informed) on site. A story which may have some substance in fact is that Ladysmith at its busiest period requested a replacement loco from Walkden once a month!

Repulse would no doubt have suffered a similar fate to the long line of scrapped predecessors had it not been taken out of service in March 1975 when the washery itself closed. The Lakeside & Haverthwaite Railway spotted an opportunity, receiving the loco on 22 September 1976 as *11 Repulse*, reverting to *Repulse* by April 2009. It can be seen in all its Austerity glory hauling passengers to this day (see colour section).

Respite

Close to Ridyard St Walkden, *c.* 1956, Hunslet Austerity *Respite* (3696/1950), heading north towards the Grosvenor Rd level crossing and Ashtons Field Colliery. The second, third and fourth wagons are of Lancashire Electric Power origin. (John Philips)

Respite storms past the Walkden sheds on 11 November 1962, *Westwood* parked on the right during its period of storage. *Respite* was based at Walkden for sixteen years, from 1952 until 1968. (P. Eckersley, Brian Wharmby)

Respite being watered at Astley Green Colliery on 10 May 1969, brakesman Ernie Gregory on top. The loco had a capacity of 1,200 gallons. (Steve Leyland).

Respite at the watering point at Astley Green Colliery, 31 May 1969, this time showing the old Lancashire boiler. Sixteen of these were in use at the colliery powering the winding engines, pumps, compressors and power plant. (Steve Leyland).

A brief *Respite* alongside the Walkden Yard loco shed fences on 27 March 1968. Note the Giesl ejector chimney and Kylpor exhaust system variable circular vent on the smokebox door. This enabled the smokebox airflow to be adjusted to suit the fuel in use. (Philip Hindley)

A fine study by Philip Hindley of *Respite* crossing Whitehead Hall Bridge at Astley Green Colliery on 8 April 1970, five days after the last coal had been wound. The loco ended up out of service at Bickershaw Colliery, Leigh, in 1981. In parts today at Preston Riversway Docklands (Ribble Steam Railway) awaiting restoration.

The naming of locomotive *Respite* has puzzled industrial locomotive and railway historians and will probably continue to do so as I have also been unable to progress the investigation! Of 0-6-0ST internal cylinder Austerity design (see specifications in main text earlier), this loco was built by the Hunslet Engine Co., Jack Lane, Leeds (3696/1950), one of five identical locos ordered by the NCB North Western Division. It arrived at Walkden Yard on 28 October 1950.

It was despatched to Astley Green Colliery, being based there from 1950 until mid-1952. It was next moved to Walkden Yard shed (Ellesmere Shed), serving a long stretch of sixteen years. In for repairs at Walkden mid-1955, late 1965 to early 1966 for Giesl ejector and Hunslet Engine Co. underfeed stoker conversion, returning to work on the Central Railway system. The loco was in for repairs once more in September 1967. *Respite* was based at Walkden and the Central Railways until December 1968. It was moved to Astley Green Colliery from December 1968 until around mid-1971. It was taken out of service and stored at Walkden Yard by mid-1971, then taken in for repairs in 1972. It was transferred to the dreaded Haig Colliery – Ladysmith Colliery washery line – by November 1972, where the stoker was removed. In November 1974 it ran away and was damaged, but somehow survived the ordeal. Returned to Walkden Yard for repairs in late November 1974, where under the new inventory system it received the number 63 000 440. It was sent to Bickershaw Colliery, Leigh, by November 1975, the Giesl ejector being removed. It carried on working there until taken out of service in 1981.

Respite (and *Gwyneth*) were given to the National Railway Museum, York, by the NCB in March 1981 with the intention of being a source of parts to build the replica broad gauge Great Western 4-2-2 *Iron Duke* for the GWR 150 celebrations, and did not join the collections as specimen locomotives. By 1985, parts from the locos were being used by Resco Railways Ltd, Erith, Kent, to build the replica. Boiler, cylinders and motion work of *Gwyneth* were used and a few parts from *Respite*. Two partners purchased the remains of *Respite* from the NRM at York; it had lain outside for many years and was devoid of all fittings. To remove *Respite* from York, the loco had to be dismantled due to low bridges en route; one low loader had the tank, cab and boiler, the chassis was on a second low loader. *Respite* arrived at Preston Riversway Docklands on 9 January 2005 (the Ribble Steam Railway), the first loco on site. The boiler, tank, cab, etc., were reunited in Chris Millers Boatyard.

Respite is currently (2013) at Preston in the workshops, consisting of frames & wheels and is in a queue of locos awaiting attention, but at least the intention is there to eventually get it steaming once more.

Revenge

Revenge parked at Walkden Yard, its base until 1967. This view is *c.* 1965–67, the secondary air adaptation of 1964 part visible as the three holes on the nearest side of the smokebox door. The foreground axles with deep balance weights are possibly ex-*Joseph* or *Bridgewater* (Hunslet 1456/1924, 1475/1924).

Revenge stands near the loco shed at Walkden Yard in 1966, its normal base until 1967. The chimney steam jet adaptation and extended bunker for the trial coal/coke fuel mixture of *c.* 1962 can be seen. Behind are the Walkden Yard laboratories and Bridgewater Offices. (Roger Fielding)

Revenge at Walkden shed, 24 August 1968. Note the extended bunker and buffers (later removed). Walkden Yard laboratories in the distance. (Philip Hindley)

Revenge at the Ladysmith Colliery washery site, Whitehaven, *c.* 1970. Possibly situated at the north (harbour) end of Haig Colliery. The loco underwent the full range of smoke-reducing and firing efficiency adaptations, chimney steam jets, Giesl ejector, secondary air and underfeed stoking, between 1965 and 1968.

An 0-6-0ST internal cylinder Austerity design (see specifications in main text earlier) built by the Hunslet Engine Co., Jack Lane, Leeds (3699/1950), one of five identical locos ordered by the NCB North Western Division. It arrived at Walkden Yard on 23 November 1950. It was initially based at Walkden Yard shed until 1967, calling in for repairs in mid-1954, 1957, early 1962. It is known to have been converted to burn the coal/coke mixture around this time, requiring an extended bunker. The experiment proved unsuccessful. Further modifications to *Revenge* came around mid-1964 with the secondary air system adaptation, holes being drilled into the smokebox to supply additional air to the firebox via special grade steel tubes. Further repairs on the loco were carried out in late 1967, the Giesl ejector conversion and a second Hunslet Engine Co. underfeed stoker being fitted by February 1968. It was transferred to the Haig Colliery – Ladysmith Colliery washery – at Whitehaven in February 1968, its stoker later removed. It was out of use by September 1973, its boiler being sent to Walkden Yard in November 1974. The loco's remains did not survive the ordeal, being scrapped there *c.* October 1976 by T. W. Ward Ltd. The washery itself had closed by 1975.

Robin Hood

A glimpse of around 1960 of the ex-Clifton & Kersley Coal Co. Ltd *Robin Hood*, Chapman & Furneaux, Gateshead, Tyneside (1200/1901), at Walkden Yard. At Robin Hood sidings, north of Swinton, until November 1957. Back to Walkden, being scrapped *c.* late 1961 to early 1962.

This outside cylinder 0-4-0 saddle tank (1200/1901) was built by Chapman & Furneaux of Gateshead, Tyneside, who had taken over the business of Black, Hawthorn (1865–1896). Chapman & Furneaux only produced a further seventy locomotives before closing in 1902, their drawings, patterns, etc., being bought by R. & W. Hawthorn & Leslie of Newcastle.

With a coupled wheelbase of 5ft 6in, 38in driving wheels and cylinders of 14in x 19in, a precisely stated tractive effort of 10,300 lbs was specified at a working pressure of 140lbs, working weight being 23 tons. Purchased originally by Clifton & Kersley Coal Co. Ltd, based north of Swinton. The loco was transferred to the former Thomas Fletcher & Sons Ltd, Outwood Colliery, after its acquisition by C. & K. C. Co. Ltd in 1909. The new company was titled the Outwood Colliery Co. Ltd

The loco joined Manchester Collieries stock in March 1929, remaining in the Robin Hood sidings and Outwood coal washery area. Only two short entries have survived among the Walkden Yard records documenting repairs carried out on *Robin Hood* in 1934 and 1935;

22 September 1934
4 new crosshead slippers fitted
2 new crosshead pins fitted
4 slide bars built up and machined
Little end brasses re bored
2 steel bushes turned and fitted in crosshead
23 copper tubes (second hand) 9ft 1 ½in long put in loco at Outwood loco shed

16 May 1935
Tyres turned up
4 new axle box brasses and axle boxes lined up
New slide rod brasses, big and little end brasses
4 new brass liners for crosshead slippers
2 new crosshead pins
1 new piston rod and other round ground up
4 new piston rings
2 valve spindles
All motion pins renewed and case hardened, all pin holes ground true
4 new spring pillars and 8 brackets for same rebushed
4 bearing springs reset
4 slide bars machined up
4 eccentric straps closed
4 new brake blocks and pins for same
1 new bunker
Boiler fittings repaired, repacked valves ground in
1 new window for cab
4 new plates in smoke box
Completed 20 December 1935

After coal industry nationalisation in January 1947, *Robin Hood* joined a number of locos which could be sent to various sites as required, arriving at Walkden Yard on 22 March 1947. The 1947 NCB loco survey stated that it had been shunting at Robin Hood sidings, Newtown, and was in poor condition.

Post-1947 and after repairs it could be found at various sites in the Manchester area ranging from Outwood Colliery, Radcliffe, and Wheatsheaf Colliery, Pendlebury, to Ashton Moss and Bradford collieries, east of Manchester. It is known to have returned to Robin Hood sidings in 1949, 1950, 1951 and 1952, filling in for loco *Kearsley*. It returned to Robin Hood sidings in mid-1954 for a lengthier spell. In August 1955 it made its way to Walkden Yard for repairs via Linnyshaw Moss, returning in November 1956. It is recorded still shunting at Robin Hood sidings in September 1957 but all work had come to an end there in November of that year. It returned to Walkden and was scrapped between December 1961 and early 1962.

Shakerley

Shakerley, Hunslet Engine Co., Leeds (736/1901), at Cleworth Hall Colliery, Tyldesley. Coupled wheelbase 10ft 9 ½in, 37in driving wheels, cylinders 14in x 18in, tractive effort 10,000 lbs at 140 lbs, working weight 23 ¾ tons. Based at Cleworth from late 1947, scrapped there in January 1962. Similar to *Edith* (See earlier).

1 Manchester Collieries cap/lapel badge of the 1930s as worn by officials, salesmen and transport staff, including locomotive crews. The flame red echoes the colour scheme used on the company wagons. (Alan Davies)

2 1930s wooden-bodied Manchester Collieries O gauge wagon by the Leeds Model Company. The detailed lithographed printing reads: 'Repairs to Walkden Yard LMS Load 12 tons'. The company also made accessories and locomotives. (Alan Davies)

3 Early 1950s National Coal Board colliery surface safety poster. The original measures approximately 4ft x 3ft. Enough accidents, particularly with gravity-sorted wagons, were occurring for this poster to be felt necessary. The ubiquitous Austerity loco bears down on a miner taking a short cut to the pithead baths.

4 A highly devoted body of enthusiasts have for many years documented the working lives of industrial locomotives and their workplaces. Industrial railway history expert and skilled modeller Philip Hindley here recreates *Francis* (Kerr Stuart 3068/1917) in NCB yellow lined O gauge detail. (Philip Hindley)

5 Philip Hindley's O gauge model of *Katherine* (Manning Wardle 1853/1914) is in its final Manchester Collieries rebuilt form, black with red lining. Knowing the level of research detail Philip has supplied for this book I can imagine that the model will survive the most rigorous scrutiny! (Philip Hindley)

6 *Princess* (LMS 2271/NSR 2) leaves Chatterley Whitfield Mining Museum, Stoke-on-Trent (closed 1991), by low loader *c.* 1994, heading off to the Churnet Valley Railway, Cheddleton. By March 2005 the NRM Shildon had acquired it. (Alan Hall, Martyn Hearson)

7 *Princess* (LMS No. 2271/NSR 2) on display at the National Railway Museum, Shildon, 19 November 2005, in non running order. Utilising the frame of *Sir Robert*, some feel it may need to be renamed as such. Indoors by late 2013. (NRM Anthony Coulls)

8 Cab of *Princess* (LMS No. 2271/ NSR 2) at NRM Shildon in 2013, water level sight glasses missing, dual access brake handle and lengthy regulator in red. Twin Gresham & Craven injectors. Said to be very original since ex-Walkden in 1967. There are rumours (late 2013) of the loco being restored to running order. (NRM Anthony Coulls)

9 *Bold* (0-4-0ST Peckett, 1737/1927) arrives at Walkden Yard by Pickford's low loader from Ravenhead Colliery, St Helens, Sunday 1 August 1965. Walkden later had an urgent request for *Bold*'s return. *Westwood*, which had been stored since 1962, was quickly patched up and sent to Ravenhead in December 1965.

10 Walkden shed, 1 May 1966. Left is *Stanley* (HE 3302/1945), front right is *Bridgewater*, (HE 1475/1924) both recently repainted, possibly for the Open Day later that summer. The part-demolished Walkden Spinning Mill stands behind the Ellesmere Colliery headgears. (Brian Wharmby collection)

11 Open Day, Walkden Yard, 8 June 1975. Visitors suffer a short hard ride in underground man-riding carriages, hauled by battery locomotive. From the left: loco shops, mechanical department, boiler house and cable shops. An event had been held in 1966, locos *Wasp*, *Harry*, *Sir Robert*, *NSR 2* and *Repulse* on view.

12 Impressive in restoration, Wigan Coal & Iron Co. Ltd locomotive *Lindsay* of 1887 seen at Steamtown, Carnforth, on 14 August 1982. The loco is still on site although the museum closed to the public in 1997. Now (2013) the base for West Coast Railways. (Dennis Sweeney)

13 Austerity *Stanley* (Hunslet 3302/1945) east of Astley Green Colliery, 8 April 1970, five days after coal production ended. The maroon and yellow lining colour scheme arrived after 1962. Giesl-converted in February 1963, Hunslet Engine Co. underfeed stoker 'upgrade' completed by March 1968. (Philip Hindley)

14 Austerity *Respite* (Hunslet 3696/1950) at Astley Green Colliery after coal production had ended, 23 May 1970. A rare survivor of the infamous Ladysmith Colliery washery system, Whitehaven, from 1972–74. At Preston Riversway Docklands (Ribble Steam Railway) since 9 January 2005. (Philip Hindley)

15 Austerities *Gwyneth*, Robert Stephenson & Hawthorns (7135/1944, WD 75185), and *Warrior*, Hunslet (3823/1954), on duty at Bickershaw Colliery, Leigh, 13 September 1977, the large coal washery behind. *Warrior* survives complete, *Gwyneth* as parts for the *Iron Duke* replica at the NRM. (Les Tindall)

16 Austerity *Warrior* Hunslet (3823/1954) at Bickershaw Colliery shed, 13 September 1977. Bunker now numbered 63 000 432. Currently (2014) with the Dean Forest Railway Society awaiting restoration. (Les Tindall)

17 Austerity *Warspite*, Hunslet (3778/1952), minus its nameplate, endures duties in the harsh environment of the Ladysmith coal washery, south Whitehaven (closed March 1975), on 9 June 1973. It was scrapped on site by T. W. Ward Ltd after August 1976. (Les Tindall)

18 The mineral line beneath the A6 at Walkden viewed to the north. Note the road sign 'NCB Walkden Central Workshops' directing you down Tynesbank. 24 September 1970. (Philip Hindley)

19 Fascinating comparison to the earlier photograph of 1970, this time on 1 November 1987. Walkden Yard's closure had been announced by British Coal on 13 August 1986. The old mineral line, long disused and filled in, and now an underpass road. (Philip Hindley)

20 Walkden shed *c.* 1966, taken from the footbridge. From the left, *Sir Robert* (NSR 72/LMS No. 2262) is being coaled, in the centre is *Stanley* (HE 3302/1945), far right is *Princess/North Stafford No. 2* (?/1923, LMS No. 2271) in store after being withdrawn from service.

21 Walkden loco shed seen on 1 November 1987, a shadow of its former self, in use as a mining transformer workshop/store. (Philip Hindley)

22 *Warrior* (HE 3823/1954) attacks Walkden Bank en route to Ashtons Field blending plant, 7 April 1970. Hauling some of the last of Astley Green Colliery's coal, the pit ceasing production on 3 April. Left distance is the Lancastrian Hall, far right the old Bridgewater Offices with clock tower. (Mike Taylor)

23 *Warrior* (HE 3823/1954) leaves Mosley Common Colliery sidings with nine wagons, heading north towards Walkden and Ashtons Field, May 1970. Leigh Rd, Boothstown, is in the distance. Currently in store at the Dean Forest Railway Society (Mike Taylor)

24 A classic photo composition of Austerity WD132 *Sapper* (formerly HE 3163/1944, WD 75113/HE 3885/1965, *Alison/Joseph*, 63 000 410) at Ewood Bridge, east of Helmshore, on the East Lancs Railway, 1 December 2013. Fitted with the Kylpor blast pipe chimney. (Richard Fox)

25 Austerity *Repulse* (HE 3698/1950) preserved at the Lakeside & Haverthwaite Railway, Cumbria. One of five similar locos ordered at the same time by the NCB, arriving at Walkden Yard on 16 November 1950. Giesl ejector fitted October 1962, Hunslet Engine Co. underfeed stoker by May 1967. (Dave Ingham)

Above left: 26 *Repulse* at work on the Lakeside & Haverthwaite Railway, 14 August 2012. Sent to Ladysmith coal washery, Whitehaven in 1973 (closed March 1975). Lakeside & Haverthwaite received the loco on 22 September 1976 as *11 Repulse*, reverting to *Repulse* by April 2009. (Nick Busschau)

Above right: 27 16 May 1965, *Sir Robert* (NSR 72/LMS No. 2262/Walkden 4 February 1937) photographed in an unusual location, having just passed under the A6 heading towards Linnyshaw Moss. Beneath the NCB NW Division logo someone has scraped the word 'DIRTY'. (Brian Wharmby collection)

28 3 April 1968, *Sir Robert* (NSR 72/LMS No. 2262/Walkden 4 February 1937) alongside Walkden shed minus maker's plate. Note the three secondary air holes on the smoke box side and also the feed to the chimney steam jet ring. Scrapped September 1969. (Brian Wharmby collection)

29 Austerity *Whiston* (HE 3694/1950) looking magnificent on Foxfield Bank, Foxfield Colliery Railway, Dilhorne, Staffordshire. Ex-Cronton and Bold Collieries, St Helens. Out of service by early 1983 and despatched to the Foxfield Railway on 26 March 1983. (John Stein)

30 1 May 1966. A 'cosmetic' paint job has been carried out on Austerity *Harry*, Hudswell Clarke (1776/1944, ex WD 71499), at Walkden Yard for the forthcoming Open Day of summer 1966, nonetheless impressive. (Brian Wharmby collection)

31 *Harry*, Hudswell Clarke (1776/1944, ex WD 71499), at Astley Green Colliery, 24 September 1970. Coal production had ceased in April, the preparation plant operating until October, coal heading to Ashtons Field blending plant. In course of restoration by Bryn Engineering, Horwich, in 2014. (Philip Hindley)

32 The coal-covered vista of Astley Green Colliery sidings from the east on 23 May 1970. On the left is a line of NCB steel internal use wagons leading to the earlier disused coal washery, dirt conveyor crossing the canal. Centre are internal wooden wagons, weighbridge and 16-ton BR steel wagons. (Philip Hindley)

33 In the summer sun of 21 August 1967, Hunslet Austerity *Wasp* (3808/1954) crosses the bridge over the old Walkden Low Level station line, the track having been lifted a couple of years before. (Brian Wharmby)

34 21 August 1967, *Wasp* (HE 3808/1954) pauses alongside the loco shed extension at Walkden Yard before continuing its run to Walkden Bank and Ashtons Field. (Brian Wharmby)

This internal cylinder 0-6-0 saddle tank (736/1901) was built by the Hunslet Engine Co. at their Jack Lane, Hunslet, Leeds site (est. 1864). Purchased new in 1901 by William Ramsden & Sons Ltd for their Shakerley Collieries, Tyldesley (Nelson and Wellington pits), and replacing a similarly named loco by Manning Wardle & Co., Leeds, of 1874. With a coupled wheelbase of 10ft 9 ½in, 37in driving wheels and cylinders of 14in x 18in, a tractive effort of 10,000 lbs was specified at a working pressure of 140 lbs, working weight being 23 ¾ tons. It was similar to loco *Edith* (see earlier).

Shakerley Collieries did not yield to the temptation to join in the Manchester Collieries merger of March 1929 but were eventually enticed by June 1935. Actually, the Shakerley Collieries were coming to the end of their working lives; Manchester Collieries only wanted them as a paper exercise in a sense to boost their overall coal production allocation. By October 1938 the two collieries had been closed but their coal reserves and allocations were secured. Later in 1938 the loco was transferred, along with *Edith*, to the Atherton Collieries to the west. Initial duties were at Howe Bridge and Gibfield collieries. Loco *Shakerley* was transferred to Chanters Colliery, to the east of Atherton, in late 1946.

The NCB loco survey of 1947 stated that the loco was still shunting at Chanters Colliery, Atherton, in good condition and 'suitable for duty'. It is known to have spent a short while at Gin Pit, Tyldesley, before moving shortly afterwards, around late 1947 onwards, to Cleworth Hall Colliery, Tyldesley, where it worked until being scrapped in January 1962. Cleworth Hall Colliery closed in January 1963.

Sir Robert

Sir Robert Burrows christens the loco now bearing his name and recently awarded title around mid-October 1937. The loco arrived at Walkden Yard on 4 February 1937. NSR 72/LMS No. 2262 was built in 1920 at the North Staffordshire Railway's Stoke-on-Trent works. Walkden Yard manager Fred Hilton stands alongside.

October 1937, recently named *Sir Robert*, complete with Union Jack on the smokebox door, poses at Walkden Yard. Coupled wheelbase 15ft 6in, overall wheelbase 23ft 0in, 60in driving wheels, cylinders 18½in x 26in. Tractive effort 19,450 lbs at 175 lbs pressure, working weight 64 tons.

Sir Robert gets down to work in the late 1930s, its glamorous arrival long forgotten. Based most of its working life at Sandhole Colliery, east of Walkden, it was taken out of service in May 1962, its boiler being condemned in March 1963. (Glen Atkinson)

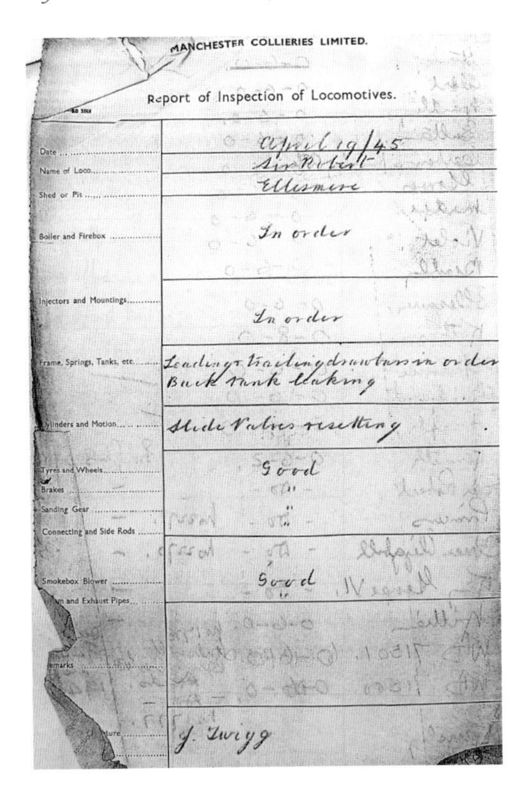

MANCHESTER COLLIERIES LIMITED.

Report of Inspection of Locomotives.

Date	April 19/45
Name of Loco	Sir Robert
Shed or Pit	Ellesmere
Boiler and Firebox	In order
Injectors and Mountings	In order
Frame, Springs, Tanks, etc.	Leading & trailing drawbars in order Back tank leaking
Cylinders and Motion	Slide Valves resetting
Tyres and Wheels	Good
Brakes	do
Sanding Gear	do
Connecting and Side Rods	do
Smokebox Blower	Good
[illegible]m and Exhaust Pipes	do
Remarks	
[illegible]ture	J. Twigg

Not the most detailed of locomotive inspections for *Sir Robert* in April 1945. Back tank leaking, slide valves needing resetting otherwise in good order. Signed by loco shop's manager, J. Twigg.

Sir Robert still looks impressive even though covered in colliery grime at Sandhole (Bridgewater) Colliery, 12 December 1950. (Alex Appleton)

Sir Robert at the loco shed at Walkden Yard, one late afternoon in 1966, firebox secondary air holes and chimney steam jet plumbing visible. (Roger Fielding)

Sir Robert heads north towards Mosley Common Colliery around early 1965. Secondary air holes can be seen on the side of the firebox, helping date the image. Leigh Road bridge, Boothstown, is in the distance, Coronation sidings are to the right of this view. The East Lancashire Road bridge would be behind the photographer.

Named after Sir Robert Burrows (1884–1964), a prominent figure in the history of the Lancashire coal mining industry. Robert came from the Fletcher/Burrows Atherton Collieries mining dynasty, son of Miles Burrows and brother to Miles Kenneth Burrows. He entered the family mining business in 1904, living close to the collieries in Atherton. Prominent in the formation of the Lancashire & Cheshire Coal Research Association, its first president, during the First World War. His interest in fuels and fuel quality ensured the development during Manchester Collieries days (post-1929) of the laboratories at Walkden Yard. Between the wars he served as chairman of Lancashire Associated Collieries and as a member of the Central Council of Colliery Owners. He had long associations with the Industrial Welfare Society and the Boys Clubs movement, as did his friend the Duke of York, later King George VI, who occasionally stayed at his home, Bonis Hall, Prestbury. He was knighted in 1937 and by 1940 had risen to High Sheriff of Cheshire. After nationalisation in 1947 he served as an unpaid NCB board member due to his specialist knowledge of fuels and fuel quality. He died in 1964.

Sir Robert (NSR 72 /LMS No. 2262/Walkden 4 February 1937) was built in 1920 at the North Staffordshire Railway's Stoke-on-Trent works. With a coupled wheelbase of 15ft 6in and overall wheelbase of 23ft 0in the loco had 60in driving wheels and inside cylinders of 18½in x 26in. A tractive effort of 19,450 lbs was specified for a working pressure of 175 lbs, working weight being 64 tons.

For most of its working life the loco was based at Sandhole (Bridgewater) Colliery, which dealt with traffic on the eastern section of the Central Railways, from Ashtons Field Colliery to Sandhole Colliery and on to Sandersons Sidings and the Worsley canal coal tip. The loco is known to have been particularly used for the heaviest workloads.

The NCB loco survey of February 1947 stated that engine work was underway on *Sir Robert* and that a new boiler would be required within two years. The entry added that the loco and the other similar examples of NSR/LMS/Midland Railway Co. 0-6-2 side tank locos, *Kenneth*, *King George VI*, *Princess* and *Queen Elizabeth*, were *'very suitable for long distance haulage and worth keeping in service'*. The loco was in Walkden for repair in January to April 1947 for new cylinders. Re-boilering of *Sir Robert* took place from late 1949 to early 1950, the boiler supplied by the Hunslet Engine Co. of Leeds. Further repairs were carried out on the loco at Walkden in late 1952, 1954, 1957, 1960 and 1961. It was taken out of service in May 1962, its boiler being condemned in March 1963.

The early 1960s saw a period of 'cannibalisation' of the 0-6-2 locos at Walkden. A more interesting example took place in early to mid-1964 when the frames, cylinders and wheels of *Sir Robert* were combined with the boiler, cab and tanks from the former *Princess/NSR 2* (now at the National Railway Museum, Shildon as *NSR 2*). The work was completed by October 1965. The loco was once more taken out of service by late 1967 to early 1968 and scrapped at Walkden Yard in September 1969. It has been commented by learned sources that the loco now at the NRM Shildon is actually *Sir Robert* as its frames form the basis for the loco *NSR 2*; the debate continues.

Spitfire

This loco was built by the Hunslet Engine Co., Jack Lane, Leeds, to their standard Austerity design (3831/1955) and delivered to Parsonage Colliery, Leigh, on 20 September 1955. By December 1958 it was to be found at nearby Bickershaw Colliery, returning to Parsonage Colliery by October 1959, then back at Bickershaw Colliery by January 1960. In November 1964, it paid a visit to Walkden Yard, receiving a Giesl ejector by 1966. Returned to Bickershaw Colliery by May 1966, it survived without major incident until being scrapped there by Mee & Cocker in March 1976.

Stanley

Austerity *Stanley* seen on a gloomy winter's day, 22 September 1964, at Walkden. Hunslet Engine Co. (3302/1945), delivered to Walkden Yard in August 1945. Giesl-converted in February 1963. (Charlie Verrall)

Stanley after overhaul at Walkden Yard, 27 September 1969. (Steve Leyland)

Stanley near the loco sheds at Walkden Yard, *c.* 1969. Around this time the loco was used to put wagons into and from connections to the ex-L&Y and LNWR at Walkden, also providing banking assistance to Ashtons Field Colliery when required.

Stanley at the eastern end of the Astley Green Colliery site on 8 April 1970, five days after the final coal had been produced at the pit. By 15 January 1971 the loco was to be found at Ladysmith coal washery, Whitehaven, being scrapped there in June 1975. (Philip Hindley)

Stanley rests at Walkden shed, 19 September 1969. The old NCB North Western Division crest is still on the cab side even though the area was now (after April 1967) within the NCB North Western Area. (Philip Hindley)

Named after Stanley Greenhalgh, one of the senior Walkden Yard engineers.

This loco was built by the Hunslet Engine Co. to their standard Austerity design (3302/1945) after the war had ended, and was delivered direct to Manchester Collieries at Walkden Yard in August 1945.

The NCB loco survey of February 1947 stated that the loco was in good condition, and along with the other Austerity locos 'doing excellent work'. Mainly used at Brackley Colliery, Little Hulton, until November 1961 when it was back at Walkden Yard for repairs, then known to have worked at Chanters Colliery, Atherton, for a short while. Back working on the Central Railways by July 1962. It underwent the Giesl conversion in February 1963 at Walkden Yard and the Hunslet Engine Co. underfeed stoker 'upgrade' was completed by March 1968. The loco was back into Walkden once more in late 1969 for overhaul.

Joe Cunliffe recalled (in 1993) that *Stanley* was in later life used to put wagons into and from the connections to the former L&Y and LNWR at Walkden and to provide banking assistance to Ashtons Field Colliery as required.

Stanley was despatched by 15 January 1971 to the 'loco death row' of Ladysmith coal washery, Whitehaven, where it survived a creditable four years, being scrapped in June 1975, shortly after the washery had closed.

Sultan

Sultan, Sharp Stewart & Co. Ltd, Atlas Works, Manchester (2490/1875), seen near Worsley *c.* 1890. Similar to *Alert* (2714/1877), *Atlas* (2909/1880), *Gower* (3090/1882) and *Wardley* (3423/1888). 48in driving wheels, cylinders 17in x 24in, boiler tubes 14ft 3in in length, working pressure 140 lbs.

An 0-6-0 internal cylinder saddle tank loco built by Sharp Stewart & Co. Ltd (2490/1875) at their Atlas works, Manchester (est. 1843). Purchased as the Bridgewater system progressed through a period of major investment and expansion. Similar to *Alert* (2714/1877), *Atlas* (2909/1880), *Gower* (3090/1882) and *Wardley* (3423/1888). Having 48in diameter driving wheels, cylinders of 17in x 24in, 132 to 135 copper boiler tubes (Walkden records) 14ft 3in in length, 2in outside diameter, 7's to 11's gauge. Working pressure was 140 lbs. The 0-6-0 SS's were of the 'long boiler' type of design, with all driving wheels ahead of the firebox. Utilising Giffard type injectors and very basic cabs open to the elements.

The Bridgewater Collieries Heavy Expenditure books at the Lancashire Record Office tell us that *Sultan* and *Alert* both received new boilers in 1907 costing £644 each (or £61,656 in 2013). *Sultan* received a new boiler once more in 1921. By the early to mid-1930s the loco was based at Sandhole Colliery, shunting at Sandersons Sidings.

The loco was scrapped in 1937–38, useable parts being retained for the remaining Sharp Stewart locos. *Sultan*'s remains were observed at Walkden Yard in August 1938, namely the cab, still with nameplate and saddle tank.

T. B. Wood

Works photograph of *T. B. Wood*, a Class H loco by Lowca Engineering Co. Ltd (233/1899), Lowca, near Whitehaven. Mainly to be found in its working life at the Tyldesley collieries: St Georges, Gin Pit and Nook Colliery. Scrapped at Walkden Yard in June 1958.

T. B. Wood at Walkden on 22 September 1957. A broken axle while shunting at Jacksons Sidings, Tyldesley, led to the Walkden journey but it was decided to scrap the loco. (P. Eckersley, Brian Wharmby)

Named after T. B. Wood, Astley & Tyldesley Collieries Company Ltd chairman (previously Astley & Tyldesley Coal & Salt Co. Ltd). Collieries: St Georges, Gin Pit and Nook Colliery, Tyldesley.

This internal cylinder 0-6-0 side tank Class H loco (233/1899) was built by the Lowca Engineering Co Ltd, Lowca, north of Whitehaven, a mining village. Fletcher & Jennings took over the business of Tulk & Ley in 1857. From 1857 to 1884 nearly two hundred locomotives were built, Lowca Engineering Company Ltd then being created. In 1905 the New Lowca Engineering Company Ltd was formed, but was short-lived. Orders had fallen plus a damaging fire in 1912 led to all production ceasing, the company closing in 1927.

With a coupled wheelbase of 11ft 6in, 46in driving wheels and cylinders of 16in x 24in, a precise tractive effort of 16,028 lbs was specified at 160lbs pressure, working weight being 39 tons.

T. B. Wood worked the Astley & Tyldesley lines without major incident for over thirty years. The loco, under Manchester Collieries, is known to have worked for a short while at Chanters Colliery, Atherton, in 1938–39, standing in for *Madge*, and in general was given lighter duties.

The NCB loco survey of February 1947 stated that the loco was shunting at Jacksons Sidings, Tyldesley, and under repair. It was based at Gin Pit sheds at the time. The loco worked on the A. & T. lines until 1957, a broken driving axle putting it out of action while shunting at Jacksons Sidings, adjacent to Tyldesley Station (LNWR/LMS). It was despatched to Walkden Yard but after assessment thought not worth repairing. Scrapped at Walkden in June 1958.

Violet

The imposing *Violet* seen here near Worsley around 1908 weighed 45 1/2 tons. New to Bridgewater Collieries from the local company Nasmyth, Wilson & Co. Ltd, Bridgewater Foundry, Patricroft, Manchester (852/1908). Costing £2,300 in 1908 (or £217,810 in 2013).

Violet seen east of Chanters Colliery, Atherton, Shakerley Estate beyond around 1960. *Violet* went to Walkden Yard for repair in summer 1962, but these were abandoned and in late 1965 the loco was scrapped.

See page 42 for another view of *Violet*.

Probably named after Violet Lambton (of the Durham coal-owning family), wife of John Egerton, Viscount Brackley, later the 4th Earl of Ellesmere. For those who may have wondered, the naming of Brackley Colliery may now be more obvious. The Bridgewater Collieries Heavy Expenditure books at Lancashire Record Office tell us that loco *Violet* was purchased by Bridgewater Collieries from Nasmyth, Wilson & Co. (est. 1867), Bridgewater Foundry, Patricroft, Manchester (852/1908) two years after the identical engine *Madge*, interestingly for £1 more at £2,300 (or £ 217,810 in 2013).

Violet, an internal cylinder 0-6-0 saddle tank, had a total wheelbase of 15ft 8in, six coupled 49in driving wheels and cylinders of 17in x 24in. Tractive effort amounted to 19,251 lbs at 85 per cent of 160 lbs boiler pressure. The boiler measured 11ft long x 4.0875ft diameter with 172 copper tubes of 2in outside diameter.

The heating surface amounted to: tubes 1,018 sq. ft, firebox 91 sq. ft, total 1,109 sq. ft. Injectors comprised two of the Gresham and Craven combination design. The blast pipe nozzle was of 4.5 inches diameter, water capacity 1,200 gallons, coal capacity 1.5 tons. Height to chimney top was 12ft 7in, length over buffers 30ft 6in, width 8ft 3.5in. The loco's working weight was 45.5 tons and factor of adhesion 5.29. The loco came equipped with cast iron brake blocks, replaced with wooden ones by Bridgewater Collieries.

The maintenance records show *Violet* was overhauled at Walkden Yard in 1923 when a new boiler was installed. Under Manchester Collieries, in June 1937, *Violet* was transferred to the Gin Pit railways, Tyldesley, the sister engine *Madge* also being transferred in March 1938, destined for Chanters Colliery, Atherton. *Violet* itself ended up at Chanters Colliery, being transferred there from Gin Pit *c.* 1943–44. *Madge* was dismantled at Gin Pit workshops in 1946, the parts being retained as spares for *Violet*.

The NCB loco survey of February 1947 stated that *Violet* was shunting at Chanters Colliery, that it was in fair condition and that a major overhaul was due.

Area reorganisation in January 1952 saw Chanters Colliery coming under the new No. 2 (Wigan) Area. As such, it was overhauled at Kirkless workshops, Aspull, in the late 1950s, being noted in the Wigan area paint scheme in August 1959.

Further area reorganisation of the coalfield took place on 1 January 1961 on creation of the West, East and North Wales areas, Chanters Colliery coming under the East area and Walkden Yard's maintenance 'umbrella'. *Violet* was despatched to Walkden Yard for repair in summer 1962 in a poor state of repair. Those repairs were abandoned part way through, the remains parked up until late 1965 when it was scrapped.

Wardley

Wardley, Sharp Stewart & Co. Ltd, Atlas Works, Manchester (3423/1888), seen at Walkden Yard *c.* 1900. 48in driving wheels, 12ft 0in coupled wheelbase, cylinders 17in x 24in, boiler tubes 14ft 3in in length, working pressure 130 lbs, tractive effort 14,100 lbs. Loaded weight 42 tons.

Wardley in the 1930s, sporting a cab of a sorts. Based at Walkden Yard until at least 1938, then at Sandhole Colliery, working Sandersons Sidings and Worsley canal coal tip, until taken out of service in 1950.

Wardley in a fascinating study by Alex Appleton of 1938–39, Walkden Yard wagon shop behind. A good clear view of the open to elements early Giffard type injectors. Scrapped at Walkden Yard in June 1953. Behind the cab is a barge coal box in for repair.

An 0-6-0 internal cylinder saddle tank loco built by Sharp, Stewart & Co. Ltd, Atlas Works, Manchester (3423/1888), the last locomotive to be built there before production was transferred to Glasgow. It was purchased as the Bridgewater system gradually moved through a period of major investment and expansion. The process had started with the purchase of *Sultan* (2490/1875), then *Alert* (2714/1877), *Atlas* (2909/1880) and *Gower* (3090/1882).

With 48in diameter driving wheels, 12ft 0in coupled wheelbase, cylinders of 17in x 24in, 132 to 135 (Walkden records estimate) copper boiler tubes 14ft 3in in length, 2in outside diameter, 7's to 11's gauge. Working pressure was 130 lbs, tractive effort 14,100 lbs. Loaded weight came to 42 tons.

As with the other 0-6-0 SS's, *Wardley* was of the 'long boiler' type of design, all the driving wheels ahead of the firebox with distinctive Giffard type injectors and very basic cabs open to the elements. The Heavy Expenditure books at the Lancashire Record Office tell us that *Wardley* received a new boiler in 1911 costing £633 (or £58,647 in 2013). Repairs are known to have been carried out on the loco at its Walkden base in 1922 when new cylinders were required. Further work was carried out in 1932 and 1935, also in late 1938 and an overhaul in 1944 where it received the cylinders from *Alert* (2714/1877).

From at least 1938 until being laid aside *c.* 1950, the loco was based at Sandhole Colliery shed for working at Sandersons Sidings and Worsley canal coal tip. The NCB loco survey of February 1947 stated that the loco was shunting on the Central Railways and was '*coming to the end of its useful life. Early replacement by an Austerity is desirable*'. Repairs were carried out at Walkden in early 1947, the loco returning to service by 27 April. It was scrapped at Walkden Yard in June 1953.

Warrior

Austerity *Warrior*, Hunslet Engine Co., Leeds (3823/1954), and driver, at Astley Green Colliery around late 1966. An old Lancashire boiler behind serves as a water tank; the colliery had sixteen of these in use.

Warrior at the eastern sidings of Astley Green Colliery, 22 March 1969, steam emerging from every orifice. The Giesl ejector conversion and, probably at the same time, the Hunslet Engine Co. underfeed stoker 'upgrade' had been carried out at Walkden Yard in 1966. (Steve Leyland)

Warrior powers up Walkden Bank at five minutes past nine on Saturday 4 April 1970, a fine study by Steve Leyland. It must be said, the Giesl ejector is taming the level of dense black smoke quite well.

Warrior at Ashtons Field Colliery (Little Hulton) coal blending site on Wednesday 8 April 1970, five days after coal production had ceased at Astley Green Colliery. The wide range of coal types and qualities being produced at the pit had made skilled blending a necessity. (Philip Hindley)

Warrior is pushed back into the shed by recently arrived *Western Queen* (63 000 445) at Bickershaw Colliery, Leigh, April 1979. The six wheeled 750hp Dorman 12QT-engined diesel electric *Western Queen* was built by GEC Traction (5479/1979) at Vulcan Foundry, Newton Le Willows. (Alan Davies)

This loco was built by the Hunslet Engine Co., Jack Lane, Leeds to their standard Austerity design (3823/1954) and delivered to Walkden Yard in September 1954, initially put to use on the Central Railways shunting. The Giesl ejector conversion (and probably at the same time, the Hunslet Engine Co. underfeed stoker 'upgrade') was carried out at Walkden Yard in 1966. It was retained at Walkden Yard as a shunter in October 1970 after closure of Mosley Common and Astley Green collieries and subsequently the Central Railway system. It was moved to Bickershaw Colliery, Leigh, by 18 July 1977, its Giesl chimney removed, being replaced by a unique home-made square version. It was also given the number 63 000 432

Warrior was taken out of service and headed off to the Dean Forest Railway for preservation in October 1984, at the height of the Miners' Strike. It was purchased by three members of the Dean Forest Railway Society on behalf of the Company, subscribing to loans to cover its purchase and transfer by road to Norchard, a former mining district. As of 2013, little overhaul work had been carried out but was in progress.

Warspite

Warspite, Hunslet Engine Co., Leeds (3778/1952), at Walkden Yard for repairs around April 1969, only a few months after being transferred to the punishing environment and harsh working practices of Harrington Colliery, Whitehaven. Scrapped at Ladysmith washery, Whitehaven, after August 1976.

Austerity *Warspite*. Fitted at Walkden with a Hill Bigwood underfeed mechanical stoker, October 1961. Its Giesl ejector conversion was completed by January 1964. Seen at Harrington Coal Preparation Plant, Lowca, Whitehaven, *c.* 1970–73.

This fascinating detail photograph was taken at Harrington Colliery, Whitehaven, on 17 September 1969 by Philip Hindley. It shows the exterior and interior aspects of the Hill Bigwood mechanical stoker adaptation on *Warspite*. Behind the hinged door on the bunker was a diesel engine driving the screw stoker. The hole in the bunker allowed for hand starting the diesel engine driving the stoker and as an intake for the cooling air flow. The exhaust from the engine would be discharged through the rectangular grille above the top edge of the bunker. The controls for the stoker can just be seen at the back of the cab. Being screw fed, the arrangement tended to jam unless the right grade/size of coal was used. The men must have found them very frustrating to use, with regular breakdowns. It is known that a total of nine NCB locos were fitted with the system but most were later removed.

Named after the famous battleship which served in both the First and Second World Wars, and which was scrapped in 1950. This loco was built by the Hunslet Engine Co., Jack Lane, Leeds, to their standard Austerity design (3778/1952) and delivered to Walkden Yard in November 1952, being based there for work on the Central Railways. In October 1961 it was fitted at Walkden with a Hill Bigwood underfeed mechanical stoker. The Giesl ejector conversion was completed by January 1964.

The loco was transferred to Harrington Colliery, Lowca village, north Whitehaven, on 8 November 1968. Back to Walkden Yard for repairs by April 1969. Back at Harrington Colliery by June 1969. Railway historian Steve Oakden states that by 1970 the mechanical stoker on *Warspite* had seized up and was not used, a fate which appears to have been common. Being open to the elements and subject to both heat and the damp, salty coastal air cannot have helped in this particular situation.

Warspite was transferred to the Ladysmith coal washery line, south Whitehaven (closed March 1975), by 17 May 1973 after Harrington coal washery had closed. It was scrapped at Ladysmith by T. W. Ward Ltd after August 1976.

Wasp

Short-lived Austerity *Wasp*, Hunslet Engine Co., Leeds (3808/1954), at its base of the time, Walkden shed, Boxing Day 1960. (P. Eckersley, Brian Wharmby)

Wasp, near Walkden Yard, *c.* 1962. Delivered to Walkden Yard January 1954. Transferred to the Gin Pit, Tyldesley, railways by August 1959. Working on the Walkden Central Railways by May 1960. Fitted with a Giesl ejector at Walkden by February 1963.

Wasp at Walkden shed *c.* 1967. By December 1963 a replacement boiler with Hunslet gas producer combustion system had been fitted (note the chimney). The very short fifteen-year working life of *Wasp* ended with scrapping at Walkden Yard by Maudland Metals of Preston in March 1969.

Built by the Hunslet Engine Co., Jack Lane, Leeds, to their standard Austerity design (3808/1954) and delivered to Walkden Yard in January 1954. It was to have a rather unexciting and short life. Initially based at Walkden Yard and working on the Central Railways, it was transferred to the Gin Pit, Tyldesley, railways by August 1959. Working on the Walkden Central Railways by May 1960, being fitted with a Giesl ejector at Walkden by February 1963. This was to be a short lived modification as by December 1963 a replacement boiler with Hunslet Engine Co. gas producer combustion system had been fitted. The old boiler from *Wasp* was fitted to *Charles* (HC 1778/1944) in January 1964. The short life of *Wasp* ended with scrapping at Walkden Yard by Maudland Metals of Preston in March 1969.

Westwood

Newly overhauled *Westwood* at Walkden around September 1960. An outside cylinder saddle tank by Hudswell Clarke & Co., Leeds (1036/1913). Ex-Platt Brothers, Moston Colliery, east Manchester. Its very varied working life ended with scrapping at Ravenhead Colliery, St Helens, in late 1968.

This outside cylinder 0-4-0 saddle tank (1036/1913) was built by Hudswell Clarke & Co. of Jack Lane, Hunslet, Leeds (est 1860). With a coupled wheelbase of 6ft 9in, 43 ¾in driving wheels and cylinders of 16in x 24in, a tractive effort of 17,300 lbs was specified at 160 lbs working pressure, working weight being 32 tons.

The NCB loco survey of February 1947 stated that the loco was working at the former Platt Brothers Moston Colliery, east Manchester. It added that an 0-6-0 type of loco was more suitable and that repairs were needed to the 'engine work'. Moston Colliery closed in June 1950, the loco was then set to work at Sandersons Sidings and Worsley canal coal tip where the loco's left hand drive arrangement found favour with crews, suiting the layout. *Westwood* then worked at various locations along with the Central Railways, including Wheatsheaf Colliery, Pendlebury, in early 1952.

In January 1955 it was despatched to Ashton Moss Colliery, east Manchester. On 30 December 1956 it was noted in Walkden Yard workshops off wheels.

From mid-1957 to September 1959 it was to be found once more at Wheatsheaf Colliery. It then returned to Walkden Yard in 1960 for repairs, emerging repainted in September. It was then transferred firstly to Astley Green Colliery then to Cleworth Hall Colliery, Tyldesley, in April 1961 for only one month. *Westwood* was next despatched to the Peel Hall Opencast Disposal Point to the north in July 1961. It returned to Walkden Yard in December 1961. Put into store at Walkden Yard by July 1962, it left Walkden Yard in November 1965 for Ravenhead Colliery, St Helens. In 1965, Walkden had had an urgent request from Ravenhead Colliery, St Helens, for loco *Bold*, still dismantled for boiler repairs. *Westwood* was rapidly patched up and despatched to Ravenhead in December 1965. It ended its relatively long and very varied working life at Ravenhead Colliery, being scrapped there by Mee & Cocker Ltd in October – November 1968.

Whiston

Austerity *Whiston* (named so after 1962), Hunslet (3694/1950). Seen moving colliery roof supports at Cronton Colliery, Halsnead, Merseyside, 22 May 1957. In the background are the closely spaced Nos 1, 2 and 3 pits. Transferred to Bold Colliery, St Helens, by January 1962, where it received the name *Whiston*.

Whiston at Bold Colliery shed *c.* late 1982. After closure of NCB Haydock Workshops in March 1963, maintenance on *Whiston* was undertaken by Walkden Yard. Out of service by early 1983, despatched 26 March 1983 to the Foxfield Colliery Railway, Dilhorne, Staffordshire. At work at the time of writing (2013).

This loco was built by the Hunslet Engine Co., Jack Lane, Leeds, to their standard Austerity design (3694/1950). Delivered in October 1950 to NCB Haydock Workshops.

By January 1951, it was at work at Cronton Colliery, Halsnead, St Helens. In for repairs at Haydock Workshops in March 1957. Despatched to Bold Colliery, St Helens, by January 1962, newly named *Whiston*. After the closure of NCB Kirkless Workshops, Wigan, in November 1962 and Haydock Workshops in March 1963, maintenance on *Whiston* was undertaken by Walkden Yard, hence inclusion in this book.

Taken out of service by early 1983, it was despatched on 26 March 1983 to the Foxfield Colliery Railway, Dilhorne, Staffordshire, where today (2013) the loco can be seen at work in all its preserved splendour.

W.H.R.

Austerity *W. H. R.* (7174/1944) was built by Robert Stephenson & Hawthorns as WD 71520 and despatched to France in May 1945. Sold by the War Department to the NCB, arriving at Walkden Yard in its WD livery on 18 June 1947. Seen around 1965 at Walkden, a glimpse of *Francis* (KS 3068/1917) on the left.

W. H. R. at Walkden shed in 1966, the secondary air holes visible adjacent to the smokebox door. Evidence of serious tank cleaning and polishing! (Roger Fielding)

W. H. R. out of use at Walkden Yard on 21 October 1967. Last seen working the previous month, it was scrapped in October 1968. One of the nameplates survives in the hands of W. H. Richards' family. The ex-LMS/Atherton Collieries system loco service brake van stands behind. (Philip Hindley)

The naming of this loco no doubt puzzled many onlookers over the years! *W.H.R.* stood for William Henry Richards, colliery manager and mining agent, a man who merits a more detailed biography than most. From 1933–1937 he was the colliery manager at the Cannock Chase, Chasetown, Nos 3 and 8 Collieries. He moved to Brackley Colliery, Little Hulton (Bridgewater Collieries), on 30 September 1938. Manager at Brackley for nine months with responsibility also for the closed Ashtons Field Colliery (pumping station), north of Walkden, and the closed Ellesmere Colliery (pumping station), adjacent to Walkden Yard. By 23 June 1939, he had been appointed manager at Astley Green Colliery after an explosion and fire on Wednesday 7 June, where the manager John Hewitt, 38, had died along with four miners. W. H. Richards had been involved in rescue work at the incident. He had risen to the heights of Mining Agent to Manchester Collieries by 1946. In the following year he became the Area General Manager, NCB North West Division, from 1 January 1947 to 25 June 1951. He died in 1951.

This standard Austerity design internal cylinder 0-6-0 saddle tank (7174/1944) was built by Robert Stephenson & Hawthorns as WD 71520. It was delivered new to the Longmoor Military Railway, Hampshire, in December 1944. It was despatched to France in May 1945 and stored at Calais until returned to the UK in May 1947. It was sold by the War Department to the National Coal Board, arriving at Walkden Yard in its WD livery on 18 June 1947.

The NCB loco survey of 1947 stated that it was intended for duty between Gibfield and Chanters collieries, Atherton, and, along with *James* (RSH 7175/1944 WD 71521), which was purchased at the same time, was being reconditioned and painted. It actually ended up being transferred to the Gin Pit railways, Tyldesley, on 30 July 1947. It returned to the Central Railways in 1964–65. It was next based at Astley Green Colliery for a short while, then based at Walkden Yard shed in 1965. It returned to Astley Green Colliery in 1966 and 1967. It was taken out of service by August 1968 and scrapped at Walkden Yard by October 1968.

William

The imposing 55-ton *William* at Walkden Yard in the late 1930s to early 1940s. Built by Sharp Stewart & Co. Ltd, Glasgow (3606/1890), for the Barry Railway as their No. 46. On the Central Railways until moving on 1 March 1949 to Astley Green Colliery for three years. Scrapped Walkden, June 1952.

Named after William Green, Bridgewater Collieries Transport Manager with both rail and canal responsibilities.

This 0-6-2 inside cylinder side tank locomotive was built by Sharp, Stewart & Co. Ltd of Glasgow (3606/1890) for the Barry Railway as their No. 46. It was sent to Walkden Yard ex-GWR (after 1923, loco No. 244), despatched directly from the Swindon works. It was supplied to Walkden by J. F. Wake of Darlington in July 1939. It was to be found at work on the Central Railways by 4 September 1939, the day after war was declared. The standard run for many years for *William* was Boothsbank to Ashtons Field Colliery.

With a coupled wheelbase of 14ft 5in and overall wheelbase of 20ft 8in, it had driving wheels 51in in diameter. Cylinders were 17 ½in x 26in, working steam pressure 160 lbs, tractive effort 18,750 lbs, loaded weight 55 tons. The NCB loco survey of 1947 stated that *William* was in use chiefly shunting on the Central Railways, based at Walkden Yard, with long distance work between collieries and washeries and main line junctions. Stated to be in good condition and very suitable for heavy haulage.

Former Walkden Yard Manager Joe Cunliffe (d. 1993) wrote in February 1990 to railway historian and photographer Alex Appleton that *William* had exotic livery applied to it in early 1947 as a result of an instruction that all manner of things were to be painted and smartened up to celebrate nationalisation. It was very costly and done to a high standard, too costly to be perpetuated!

Repairs were carried out on the loco at Walkden in 1948, returning to duty by March 1949. It was transferred to Astley Green Colliery on 1 March 1949, but taken out of use and stored at Walkden Yard by May 1951. It was officially withdrawn from service on 15 October 1951 and was scrapped at Walkden in June 1952.

Witch

Austerity *Witch*, Hunslet, Leeds (3842/1956). Initially based at Walkden its first move was to Chanters Colliery, Atherton, around July 1962, where it was photographed. Back to the Central Railways by 11 September 1965. Scrapped by Maudland Metals of Preston in April 1969.

This loco was to have a surprisingly short and unexciting working life. Built by the Hunslet Engine Co., Jack Lane, Leeds, to their standard Austerity design (3842/1956) and delivered to Walkden Yard in September 1956, it was ordered at the same time as *Wizard* (HE 3843/1956). Initially based at Walkden, its first move was to Chanters Colliery, Atherton, around July 1962 as a replacement for *Violet* (Nasmyth, Wilson 852/1908). Into Walkden Yard for repairs by March 1965, then leaving to work on the Central Railways by 11 September 1965. The closure of Sandhole washery in September 1968 probably hastened its demise, being scrapped by Maudland Metals of Preston in April 1969.

Wizard

Austerity *Wizard*, Hunslet, Leeds (3843/1956), photographed shortly after delivery to Walkden in September 1956. Seen on the Ashtons Field Sandhole Colliery line, thirteen wagons visible. Note the typical Bridgewater Collieries splayed-leg power line pylons. (John Philips)

This loco was built by the Hunslet Engine Co., Jack Lane, Leeds, to their standard Austerity design (3843/1956) and delivered to Walkden Yard on 21 September 1956. It was ordered at the same time as the identical loco *Witch* (HE 3842/1956) and was to have a similar short and uneventful life.

Repairs are known to have been carried out on the loco at Walkden in mid-1961. It was normally based at Sandhole Colliery shed and later at Walkden Yard, Ellesmere Shed. Sandhole washery closed in September 1968, *Wizard* being taken out of service in December 1968. Maudland Metals of Preston bought the loco for scrap in April 1969

Wizard at Walkden shed around the mid-1960s. It had been fitted with a Giesl ejector by March 1963 and chimney steam rings by May 1964.

Steam rings and Giesl ejector temporarily overridden for the photographer of *Wizard* in the mid-1960s, probably at Sandhole Colliery. The washery at Sandhole closed in September 1968, *Wizard* being taken out of service in December. Sold for scrap in April 1969.

The Collieries

The main reason for the creation of the Bridgewater Collieries and later Manchester Collieries rail network was to transport coal and mining materials, accessing at various points the washeries, canal tips and wharves along with main line connections. All the collieries varied in age and surface layout and in a sense had a unique placing in the landscape. Here the main collieries are listed alphabetically.

Ashton Moss

The towering pit headgears of Ashton Moss Colliery, Ashton-under-Lyne, east Manchester, around 1913. Shafts visible 954 and 934 yards in depth. The colliery was sunk after 1875 and closed in September 1959, its workforce being 503 at the time.

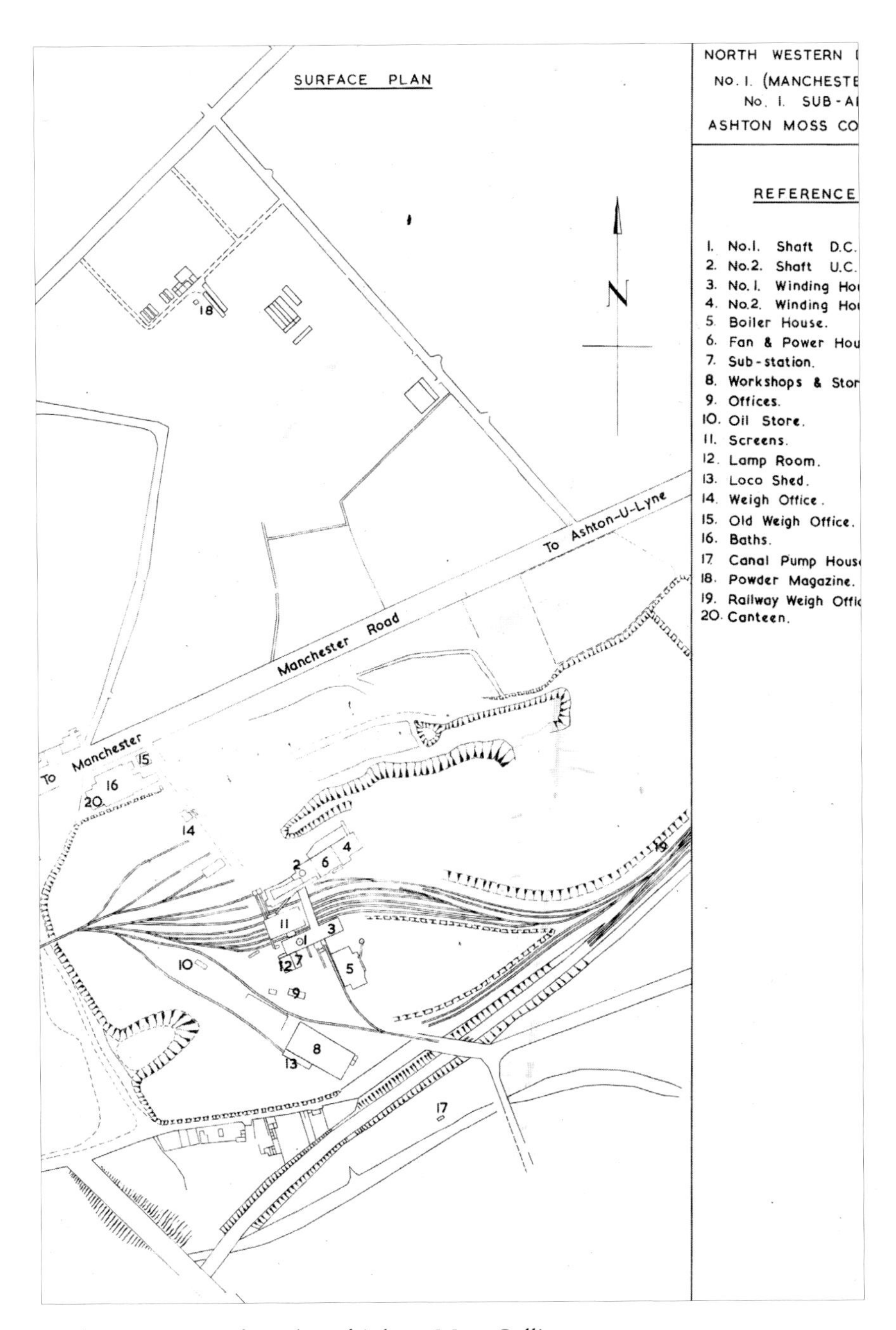

Mid-1950 NCB surface plan of Ashton Moss Colliery:
1. No. 1 shaft downcast, 2. No. 2 shaft upcast, 3. No. 1 winding house, 4. No. 2 winding house, 5. Boiler house, 6. Fan and power house, 7. Sub station, 8. Workshops and stores, 9. Offices, 10. Oil store, 11. Screens, 12. Lamp room, 13. Loco shed, 14. Weigh office, 15. Old weigh office, 16. Baths, 17. Canal pump house, 18. Powder magazine, 19. Railway weigh office, 20. Canteen

Ashtons Field

Ashtons Field Colliery No. 1 and 2 Pits, Little Hulton, *c.* 1929. Sunk *c.* 1852, 515 yards to the Arley seam, also assessing the Worsley underground canals. Closed November 1929, 800 losing their jobs.

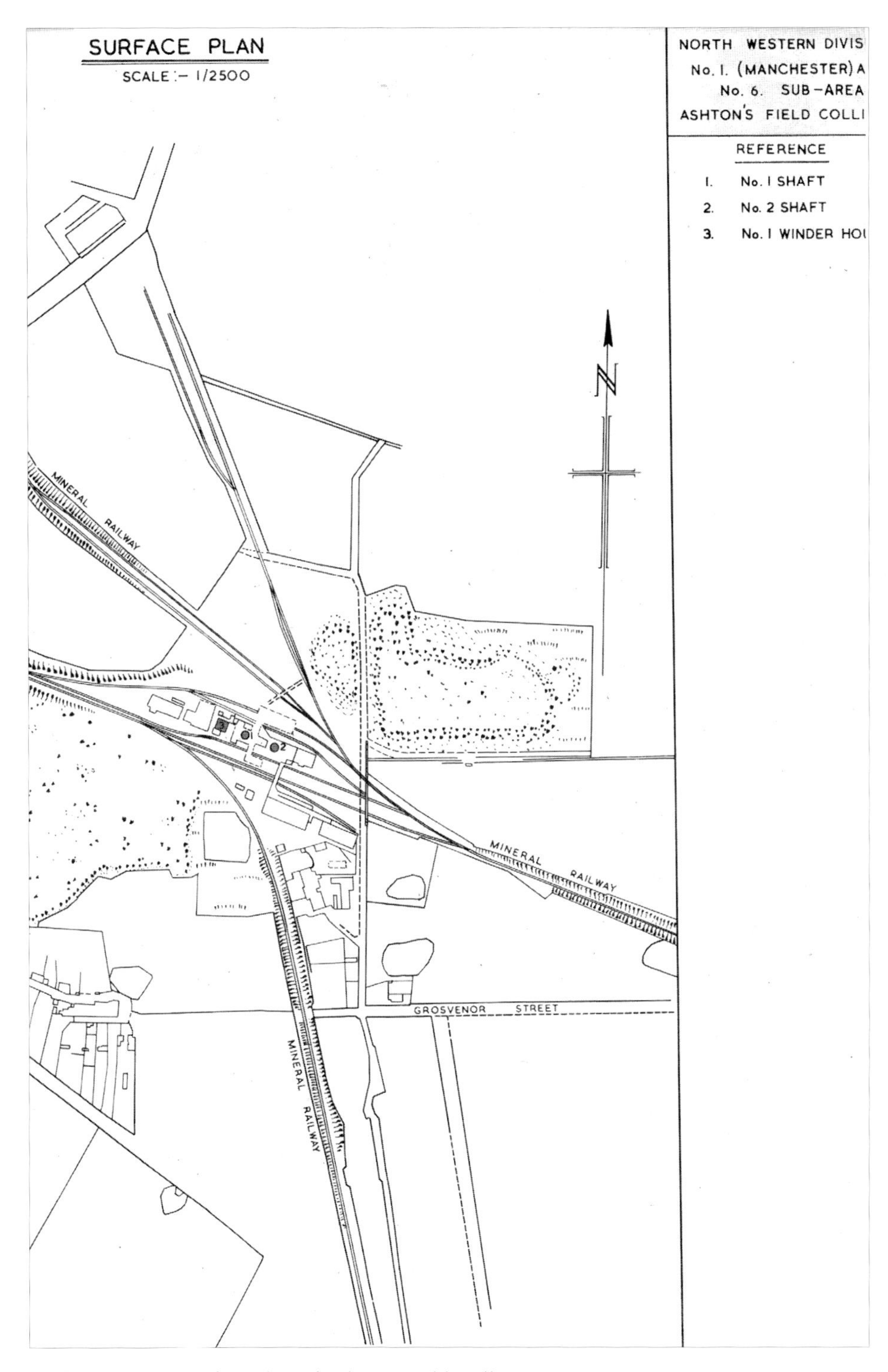

Mid-1950 NCB surface plan of Ashtons Field Colliery:
1. No. 1 downcast shaft, 2. No.2 upcast shaft, 3. No. 1 winder house. Note the Grosvenor St level crossing north of Walkden town centre.

Astley Green

Astley Green Colliery, *c.* 1932. Sunk from 1908 to 1912. West view showing the sidings and sixteen lines beneath the coal classification screens. Bridgewater Canal (Leigh Branch) to the right, Manchester Collieries 'box boats' lined up. Coal production ceased 3 April 1970, 1,382 men losing their jobs. (John Taylor)

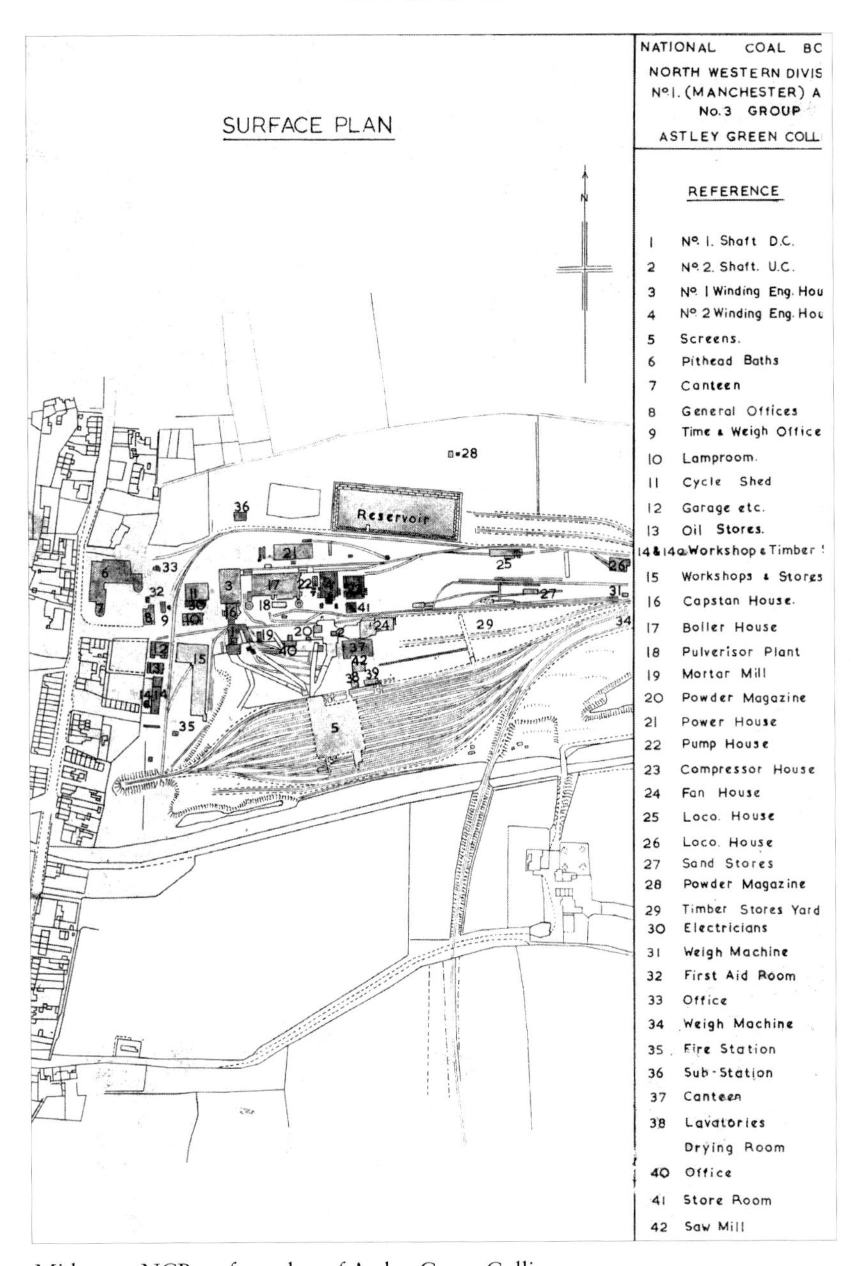

Mid-1950 NCB surface plan of Astley Green Colliery;
1. No. 1 shaft downcast, 2. No. 2 upcast shaft, 3. No. 1 winder house, 4. No. 2 winder house, 5. Screens, 6. Pithead baths, 7. Canteen, 8. Offices, 9. Time and weigh office, 10. Lamproom, 11. Cycle shed, 12. Garage, 13. Oil stores. 14/14a. Workshop and timber shed, 15. Workshops and stores, 16. Capstan house, 17. Boiler house, 18. Pulveriser plant, 19. Mortar mill, 20. Powder magazine, 21. Power house, 22. Pump house, 23. Compressor house, 24. Fan house, 25. Loco shed, 26. Loco shed, 27. Sand stores, 28. Powder magazine, 29 Timber stores yard, 30. Electricians, 31. Weigh machine, 32. First aid room, 33. Office, 34. Weigh machine, 35. Fire station, 36. Sub station, 37. Canteen, 38. Lavatories, 39. Drying room, 40. Office, 41. Store room, 42. Saw mill

Bedford

A late 1930s view from the east of John Speakman's Bedford Colliery, east of Leigh. Victoria coal classification screens in view. Sunk from 1874 onwards, closed in October 1967, 581 miners and 131 surface workers losing their jobs. The loco may be *Gower* (0-6-0 ST IC Sharp Stewart 3090/1882), on site mid-1930s to 1943.

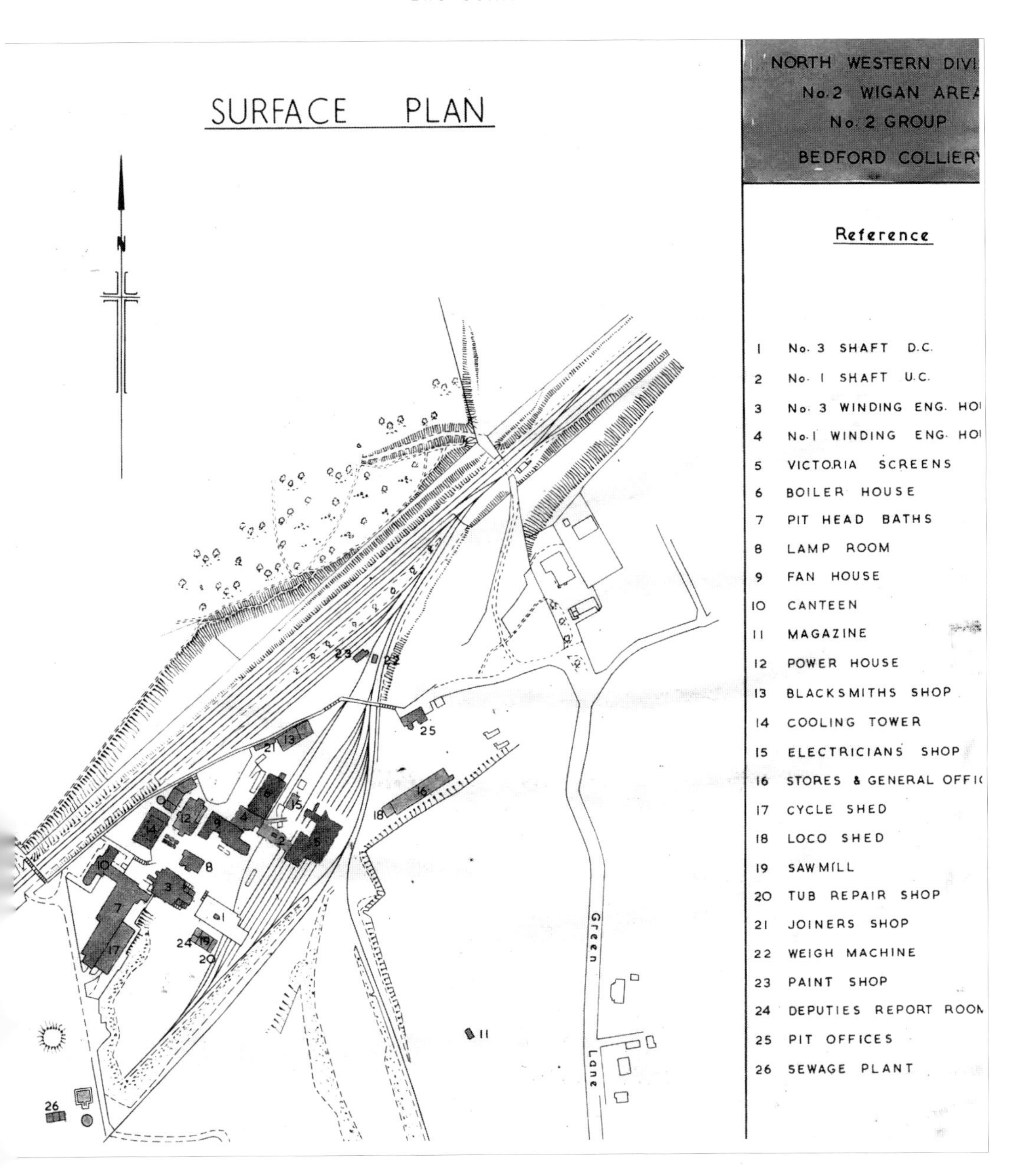

Mid-1950 NCB surface plan of Bedford Colliery:
1. No. 3 downcast shaft, 2. No. 1 upcast shaft, 3. No. 3 winder house, 4. No. 1 winder house, 5. Victoria screens, 6. Boiler house, 7. Pithead baths, 8. Lamp room, 9. Fan house, 10. Canteen, 11. Magazine, 12. Power house, 13. Blacksmiths shop, 14. Cooling tower, 15. Electricians shop, 16. Stores and office, 17. Cycle shed, 18. Loco shed, 19. Saw mill, 20. Tub repair shop, 21. Joiners shop, 22. Weigh machine, 23. Paint shop, 24. Deputies report room, 25. Offices, 26. Sewage plant

Brackley

Brackley Colliery, sunk *c.* 1858, accessed the Worsley underground canal system. Closed May 1964, this photo March 1966. 715 men lost their jobs. A loco was retained for shunting Cutacre tip, *Warspite* standing in the shed, on site till December 1966. *Witch* replaced it until late 1967, when the site was closed.

Brackley Colliery, east of Little Hulton, one of the early original Bridgewater collieries, 21 April 1951. Brackley coal was being sent up in tubs, washed, screened, loaded and used at Kearsley Power Station without being man handled. Pit tubs on the left, main line wagons below. No. 2 Pit left, No. 1 right.

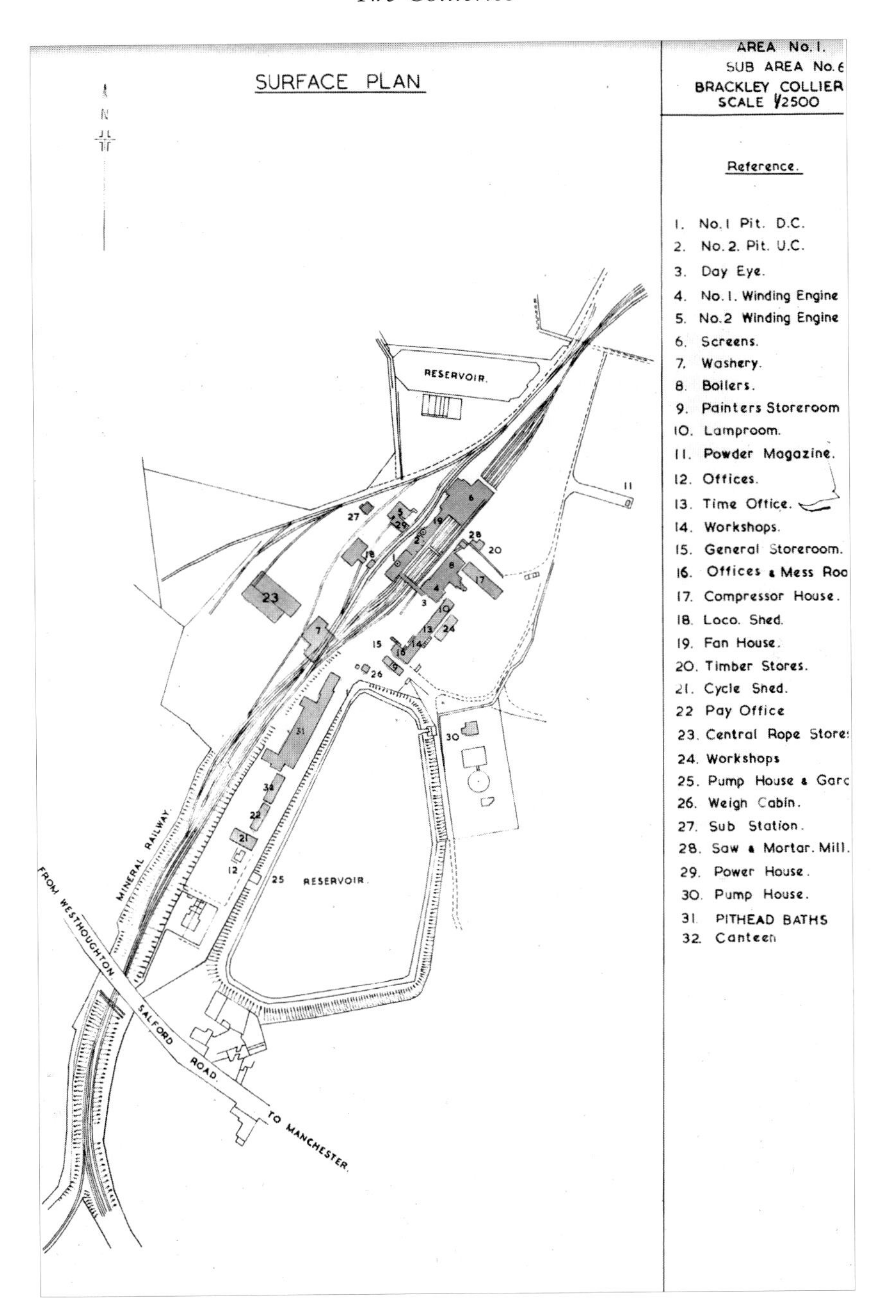

Mid-1950 NCB surface plan of Brackley Colliery:
1. No. 1 pit downcast shaft, 2. No. 2 Pit upcast shaft, 3. Day eye (drift mine entrance). 4. No. 1 winder, 5. No. 2 winder. 6. Screens, 7. Washery, 8. Boilers, 9. Painters store, 10. Lamproom, 11. Powder magazine, 12. Offices, 13. Time office, 14. Workshops, 15. General stores, 16. Offices and mess room, 17. Compressor house. 18. Loco shed, 19. Fan house, 20. Timber stores, 21. Cycle shed, 22. Pay office, 23. Rope stores, 24. Workshops, 25. Pump house and garage, 26. Weigh cabin, 27. Sub station, 28. Saw and mortar mill, 29. Power house, 30. Pump house, 31. Pithead baths, 32. Canteen

Bradford

Bradford Colliery, Bradford, east Manchester, was a truly ancient one, the death of a miner recorded in 1622. Seen here in 1935 as the colliery moved to Manchester Collieries, control. The 1928 Avonside 0-4-0 ST loco (un-named or numbered) can be seen to the left. The pit closed in September 1968, 1,556 men losing their jobs.

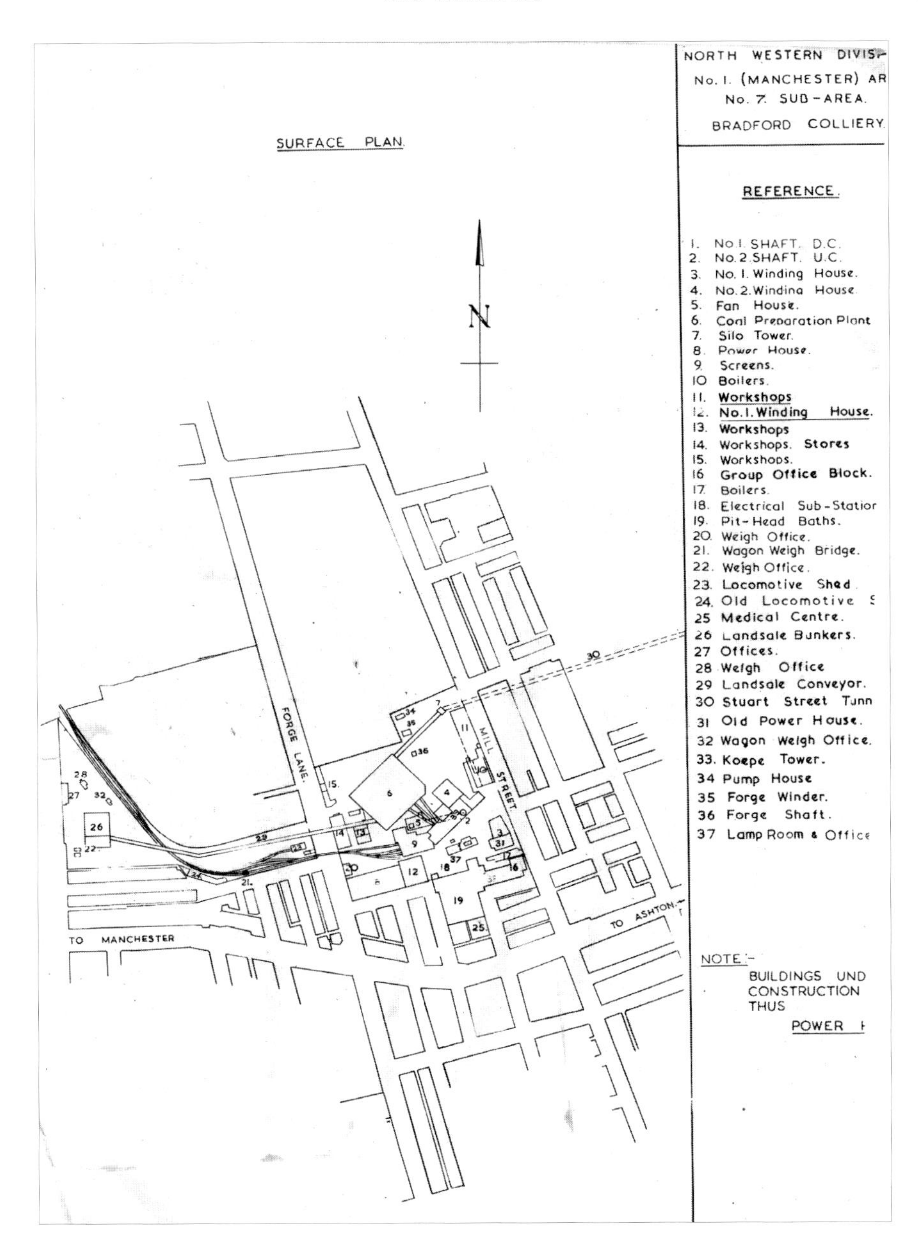

Mid-1950 NCB surface plan of Bradford Colliery, east Manchester (mid 1947–54 reconstruction);
1. No. 1 downcast shaft, 2. No. 2 upcast shaft, 3. No. 1 winder, 4. No. 2 winder, 5. Fan house, 6. Coal preparation plant, 7. Silo tower, 8. Power house, 9. Screens, 10. Boiler, 11. Workshops, 12. No.1 winder, 13. Workshops, 14. Workshops and stores, 15. Workshops, 16. Offices, 17. Boilers, 18. Sub station, 19. Pithead baths, 20. Weigh office, 21. Wagon weighbridge, 22. Weigh office, 23. Loco shed, 24. Old loco shed, 25. Medical centre, 26. Landsale bunkers, 27. Offices, 28.Weigh office, 29. Landsale conveyor, 30. Stuart St tunnel, 31. Old power house, 32. Wagon weigh office, 33. Koepe winder tower, 34. Pump house, 35. Forge winder, 36. Forge shaft, 37. Lamp room and offices

Chanters

Chanters Colliery, east of Atherton, seen *c.* 1940 from the north-east. No. 2 Pit is in view along with the boiler house, note the old wooden water cooling towers. Coal winding at No.2 Pit (in view) ceased in 1964, two years before the colliery closed.

Chanters Colliery No. 2 upcast pit headgear photographed for the Manchester Collieries magazine *Carbon* in September 1933. The colliery finally closed in June 1966, 693 men losing their jobs.

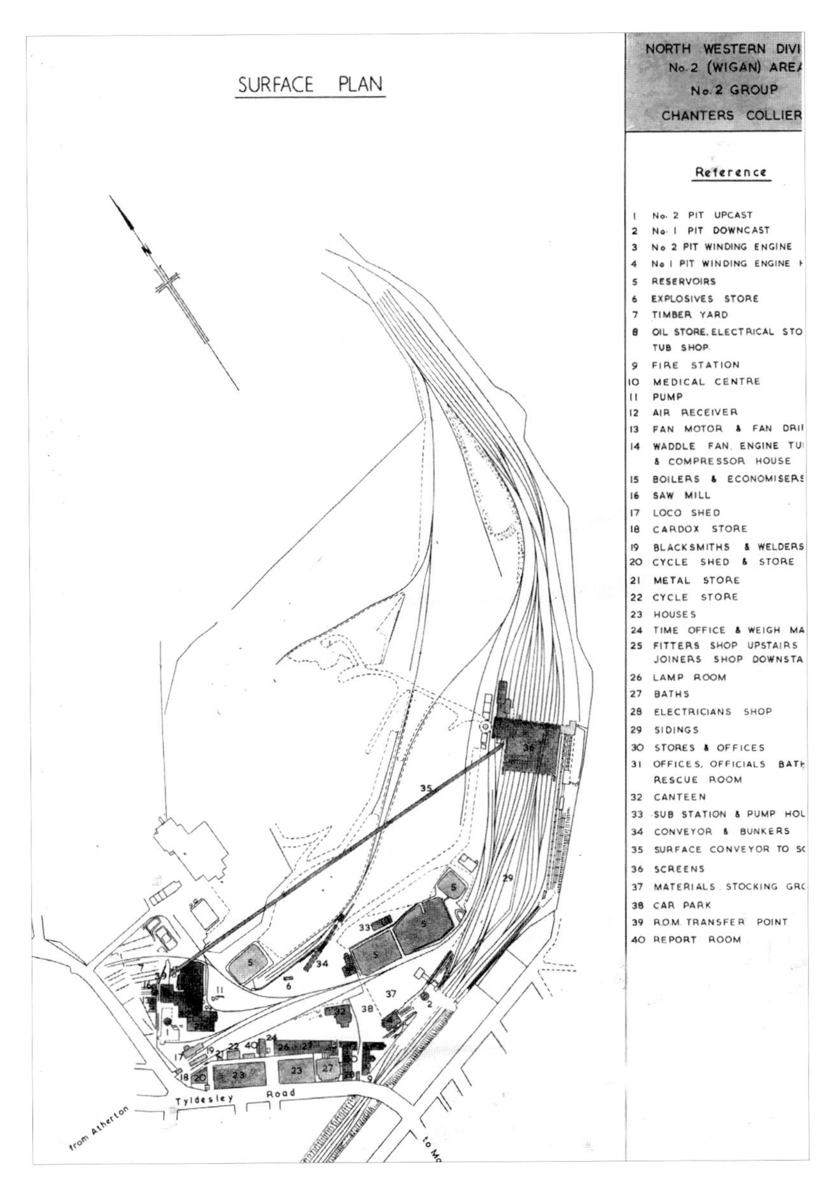

Mid-1950 NCB surface plan of Chanters Colliery, east of Atherton; 1. No. 2 upcast pit, 2. No. 1 downcast pit, 3. No. 2 downcast winder, 4. No. 1 winder, 5. Reservoirs, 6. Explosives store, 7. Timber yard, 8. Oil store, electrical store, tub shop, 9. Fire station. 10. Medical centre, 11. Pump, 12. Air receiver, 13. Fan motor and drift, 14. Waddle fan, engine and compressor house, 15. Boilers and economisers, 16. Saw mill, 17. Loco shed, 18. Cardox store, 19. Blacksmiths and welders, 20. Cycle shed and store, 21. Metal store, 22. Cycle store, 23. Houses, 24. Time office and weigh machine, 25. Fitters and joiners shop, 26. Lamp room, 27. Baths, 28. Electricians shop, 29. Sidings, 30. Stores and offices, 31. Offices, officials baths, rescue room, 32. Canteen, 33. Sub station and pump house, 34. Conveyor and bunkers, 35. Surface conveyor to screens, 36. Screens, 37. Materials stocking ground, 38. Car park, 39. Run of mine coal transfer point, 40. Report room

Cleworth Hall

Cleworth Hall Colliery, east of Tyldesley, after closure in January 1963, seen from the west. Sunk 681 yards to the Arley seam from the 1870s to 1914 by Tyldesley Coal Co. They also owned Peelwood Colliery, Combermere Colliery and Bank House Colliery, north of Tyldesley. 535 men lost their jobs on closure.

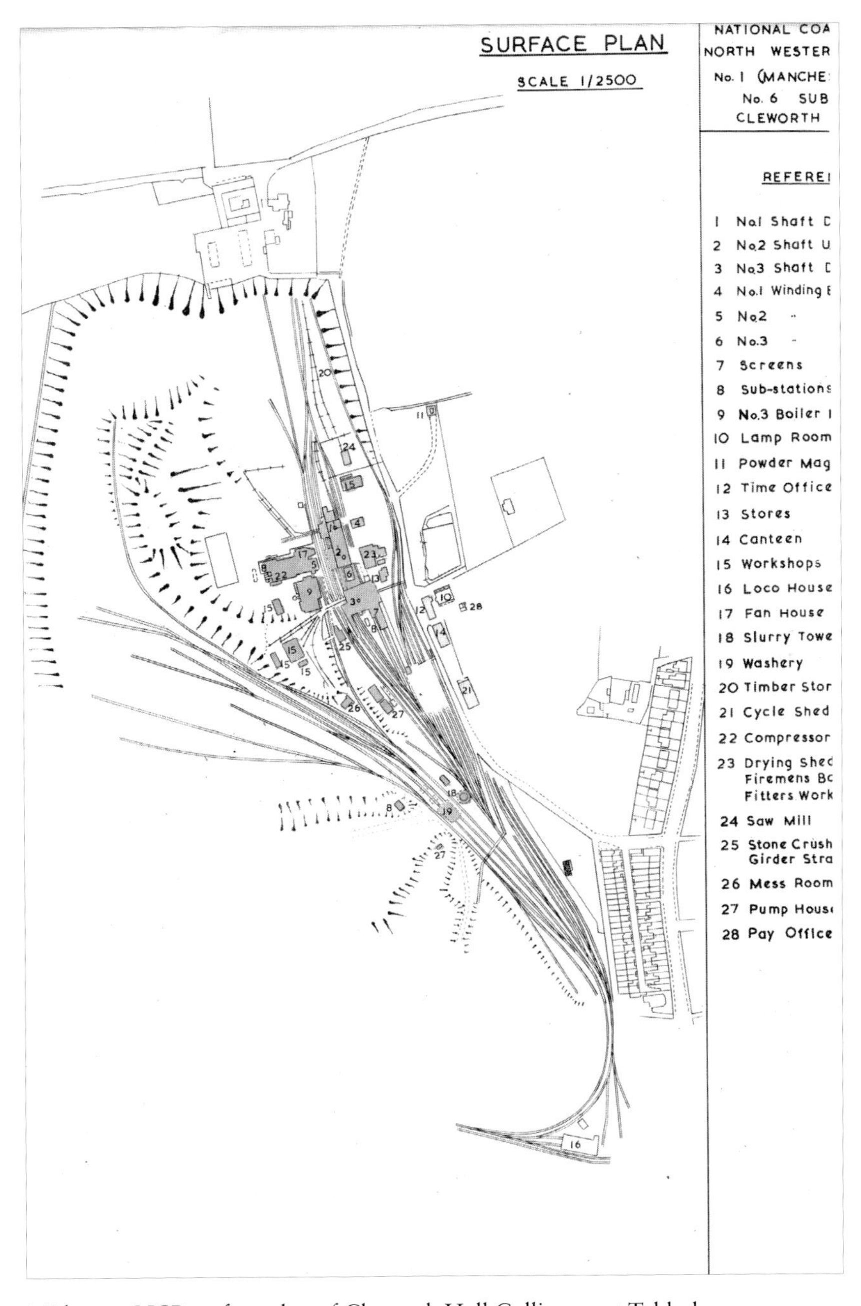

Mid-1950 NCB surface plan of Cleworth Hall Colliery, east Tyldesley:
1. No. 1 downcast shaft, 2. No. 2 upcast shaft, 3. No. 3 downcast shaft, 4. No. 1 winder, 5. No. 2 winder, 6. No. 3 winder, 7. Screens, 8. Sub stations, 9. No.3 boiler house, 10. Lamp room, 11. Powder magazine, 12. Time office, 13. Stores, 14. Canteen, 15. Workshops, 16. Loco house, 17. Fan house, 18. Slurry tower, 19. Washery, 20. Timber store, 21. Cycle shed, 22. Compressor, 23. Drying shed, firemens report room, fitters workshop, 24. Saw mill, 25. Stone crusher, girder straightener, 26. Mess room, 27. Pump house, 28. Pay office

Ellesmere

The fine engine house of Ellesmere Colliery, adjacent to Walkden Yard, sunk *c.* 1866. The shafts accessed the main underground canal. The engine house originally made use of a twin cylinder Naysmyth Wilson 30 x 50in vertical steam winding engine, electrified in 1936. The colliery ceased production in 1921, being retained as a pumping station until 1968. Sadly a preservation campaign was unsuccessful.

Gibfield

Gibfield Colliery, west Atherton seen, after closure in August 1963. Viewed from Coal Pit Lane, accessed via Wigan Rd. Sunk around 1872, older shafts on site of around the 1770s. The colliery had worked virtually all its possible reserves, Fletcher Burrows & Co. and later the NCB having worked the unit efficiently.

Recently discovered detailed view of Gibfield Colliery *c.* 1932, view east, Wigan Rd top right. First pithead baths in Britain top centre, Stephenson's Bolton to Leigh line crossing the site. Edward Ormerod's safety winding detaching hook workshop and cottage lie bottom centre alongside the lodge. (John Taylor)

Gin Pit

Gin Pit, Tyldesley, a classic colliery / community unit, seen here around 1932. Note the village bottom left, colliery (sunk *c.* 1850, closed 1955) and Gin Pit Workshops top right, also the lines and extensive sidings pointing towards St Georges Colliery, where the LNWR connection and Jacksons sidings could be found. (John Taylor)

Howe Bridge

Howe Bridge Colliery, west of Atherton, was sunk by John Fletcher & Others *c.* 1849–50 149 yards to the Seven Feet seam. Later to be owned and operated by Fletcher Burrows & Co. (formed 1892). The pit closed in September 1959, this photo was taken soon afterwards, 353 men losing their jobs.

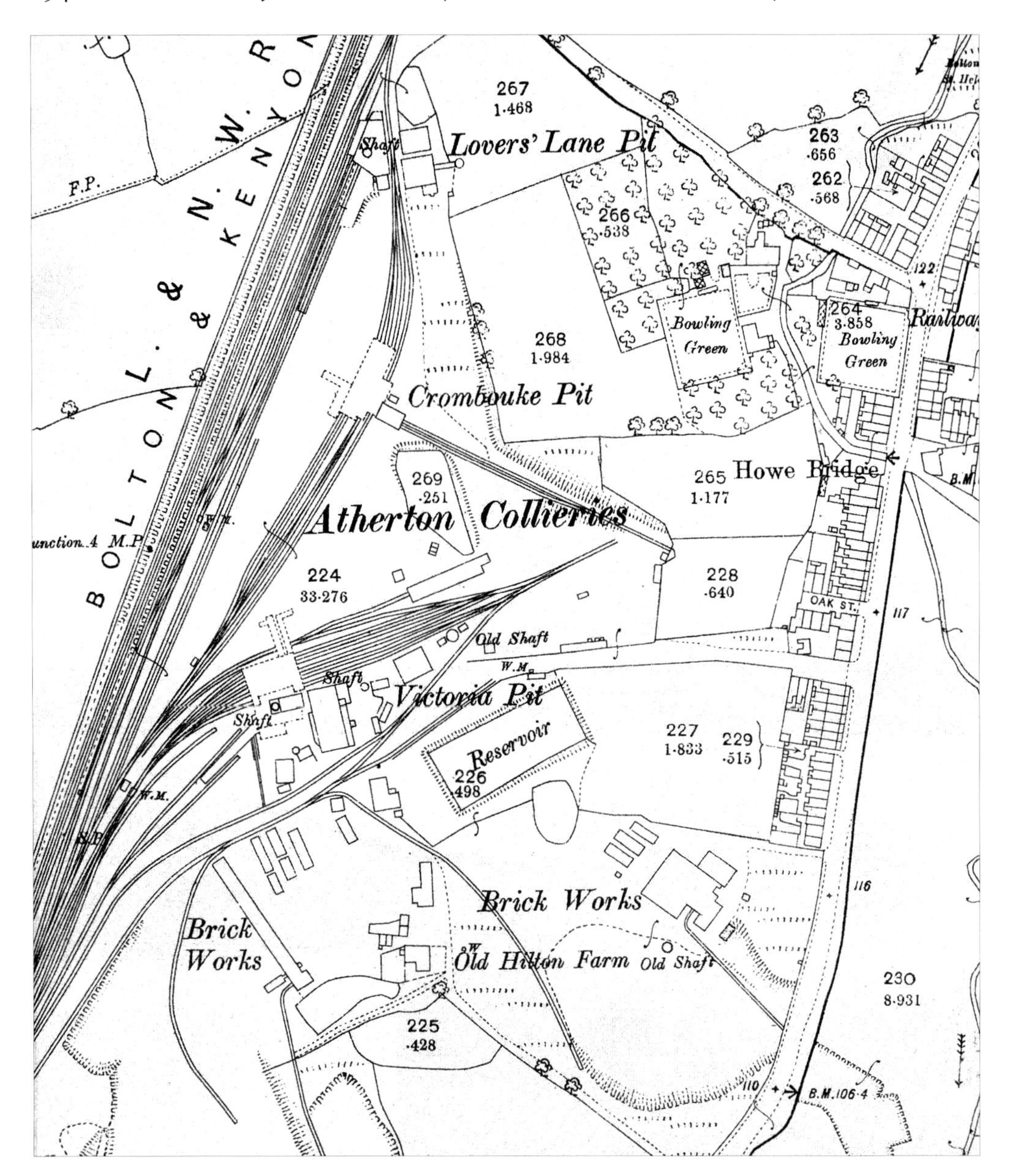

Howe Bridge Colliery, east Atherton, 1926 25in OS plan. To the left the former Bolton to Leigh Railway. Note the lines to the SLT, later LUT tramways / bus headquarters and generating station, demolished in 1998.

Linnyshaw

Bridgewater Collieries' Linnyshaw Colliery, east of Walkden *c.* 1921. Sunk *c.* 1852. Note the BC wagons and central vertical steam winding engine house serving two single cage shafts. Coal production ceased 31 March 1921, the shafts being retained as a pumping station for Sandhole (Bridgewater) Colliery until 1937.

Mosley Common

Above left: Mosley Common Colliery from the air, an RAF photograph of 1946. From the bottom edge we see the mineral line from Boothstown Canal Tip heading north below Leigh Rd then the East Lancs Rd, past the colliery, Ellenbrook Brickworks then west of Newearth Rd towards Walkden Yard.

Above right: Mosley Common Colliery oblique view of April 1951 taken for the NCB, viewed from the south-west. Compare with the RAF view of 1946. The new coal washery is under construction, to be the largest capacity washery in Britain at 600 tons per hour. The East Lancs Rd is bottom right, the slag heaps are now home to exclusive housing.

Mosley Common Colliery, east of Tyldesley was sunk from 1861 to 1885, five shafts in total accessing huge coal reserves to the south. Viewed from the east in the early 1940s are No. 2, No. 1 and No. 4 Pits. Note the lines of Manchester Collieries and Lancashire Electric Power (Kearsley Power Station) wagons. (Mrs White)

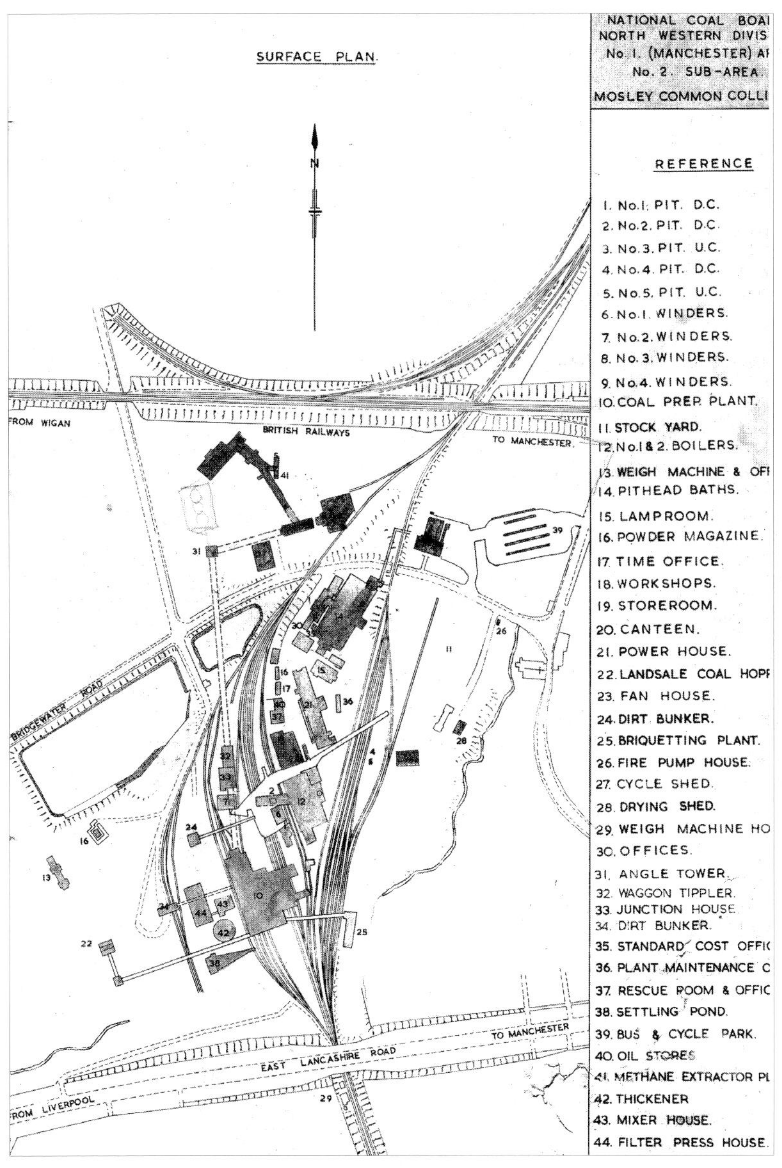

Mid-1950 NCB surface plan of Mosley Common Colliery, east Tyldesley: 1. No. 1 downcast shaft, 2. No. 2 downcast shaft, 3. No. 3 upcast shaft, 4. No. 4 downcast shaft, 5. No. 5 upcast shaft, 6. No. 1 winder, 7. No. 2 winder, 8. No. 3 winder, 9. No. 4 winder, 10. Coal preparation plant, 11. Stock yard, 12. No.1 & 2 boilers, 13. Weigh machine and offices, 14. Pithead baths, 15. Lamproom, 16. Powder magazine, 17. Time office, 18. Workshops, 19. Storeroom, 20. Canteen, 21. Power house, 22. Landsale coal hopper, 23. Fan house, 24. Dirt bunker, 25. Briquetting plant, 26. Fire pump house, 27. Cycle shed, 28. Drying shed, 29. Weigh machine house, 30. Offices, 31. Angle tower, 32. Wagon tippler, 33. Junction house, 34. Dirt bunker, 35. Standard cost office, 36. Plant maintenance office, 37. Rescue room and office, 38. Settling pond, 39. Bus and cycle park , 40. Oil stores, 41. Methane extraction plant, 42. Thickener, 43. Mixer house, 44. Filter press house.

Moston

Moston Colliery, Nuthurst Road bridge over the Manchester to Middleton L&Y railway *c.* 1900 to 1910. Mining in the area is recorded back to the sixteenth century, deep shafts on site being sunk after the 1820s, 1850s and 1880s.

Extensive reorganisation took place at Platt Brothers' Moston Colliery in 1945/6, including reinforced concrete encasing of the upcast shaft headgear in preparation for shaft deepening and skip winding introduction, but nationalisation arrived in January 1947. The NCB had other ideas and the colliery closed in June 1950. This was partly due to the difficulties of working a combination of steeply inclined and thick seams. 658 men lost their jobs on closure.

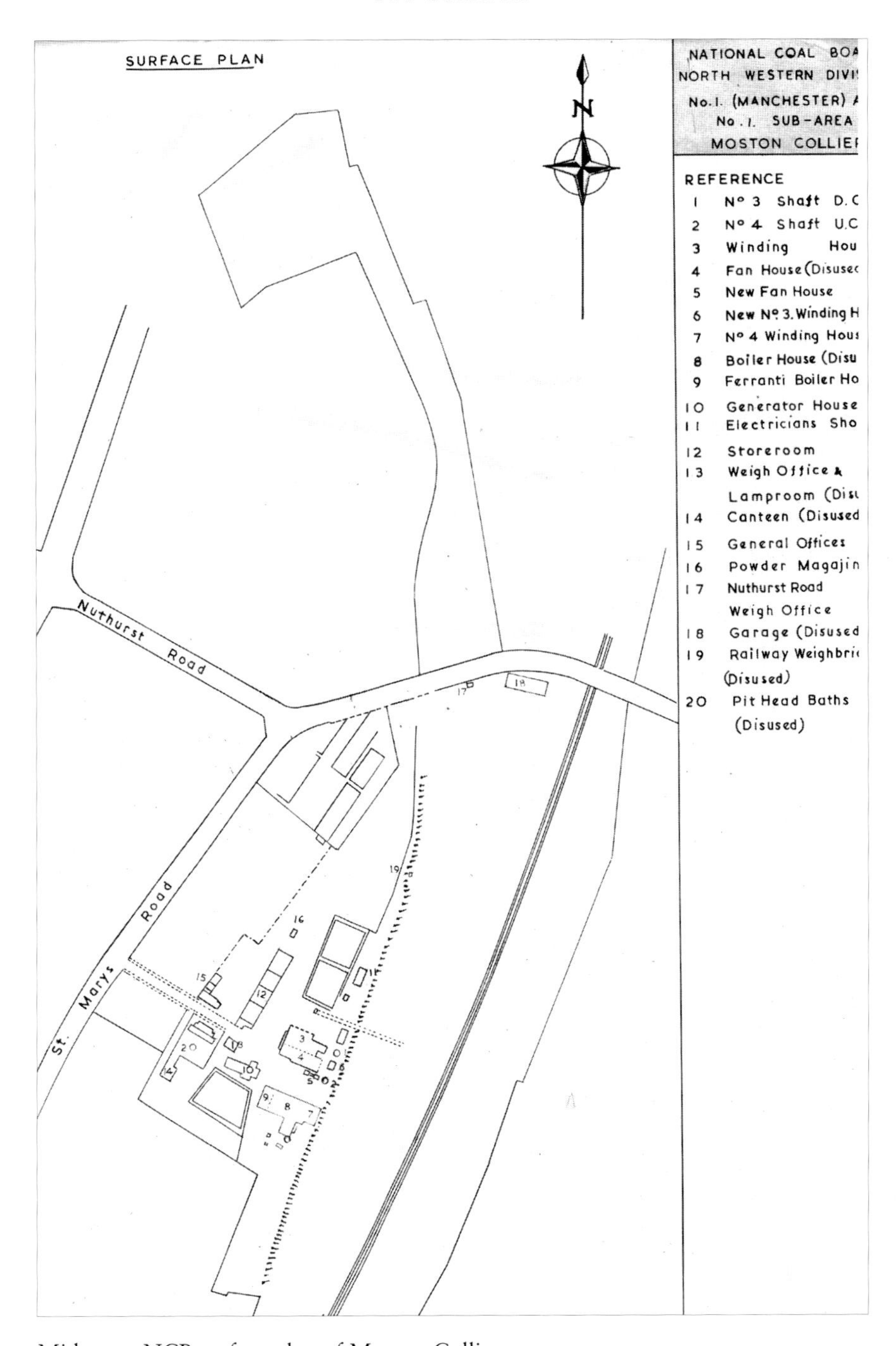

Mid-1950 NCB surface plan of Moston Colliery;
1. No. 3 downcast shaft, 2. No. 4 upcast shaft, 3. Winding engine house, 4. Fan house (disused), 5. New fan house, 6. New No. 3 winding house, 7. No. 4 winder, 8. Boiler house (disused), 9. Ferranti boiler house, 10. Generator house, 11. Electricians shop, 12. Storeroom, 13. Weigh office, 14. Canteen (disused), 15. General offices, 16. Powder magazine, 17. Nuthurst Rd weigh office, 18. Garage (disused), 19. Railway weighbridge (disused), 20. Pithead baths (disused)

Newtown

Newtown Colliery, Pendlebury, was sunk around 1875 by Clifton & Kersley Coal Co. Manchester Collieries wanted to close the pit in 1943 due to losses. The Ministry of Fuel and Power agreed on partial closure, underwriting any losses. This south west view is of 1962, the colliery closing in March 1961, 476 men losing their jobs. (Salford Local Studies Library)

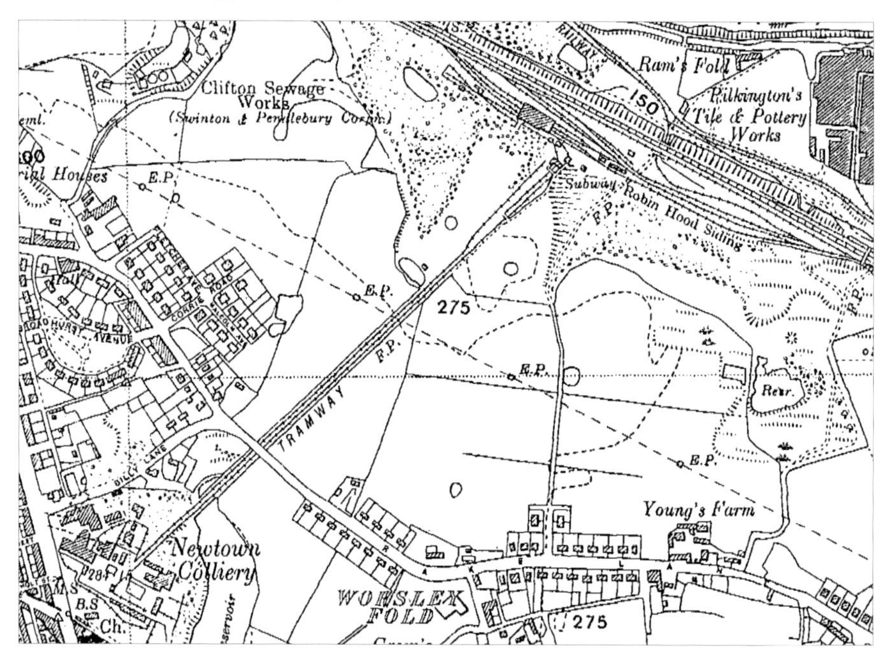

Newtown Colliery, Pendlebury, shown on the 6in to a mile 1938 OS map, lower left. Note the tub tramway which ran steeply downhill to Robin Hood Sidings and screens (the colliery did not operate a washery) which ceased operation in November 1957. Originally an endless chain system, replaced by endless rope in the 1930s.

Nook

Nook Colliery was south of Tyldesley and south of Gin Pit village. Sunk from 1848 onwards, five shafts on site eventually, the deepest No. 4 reaching 944 yards to the Arley seam. This 1930s view is from the east, spare pit cages in the foreground. The colliery's brickworks chimneys are in the distance.

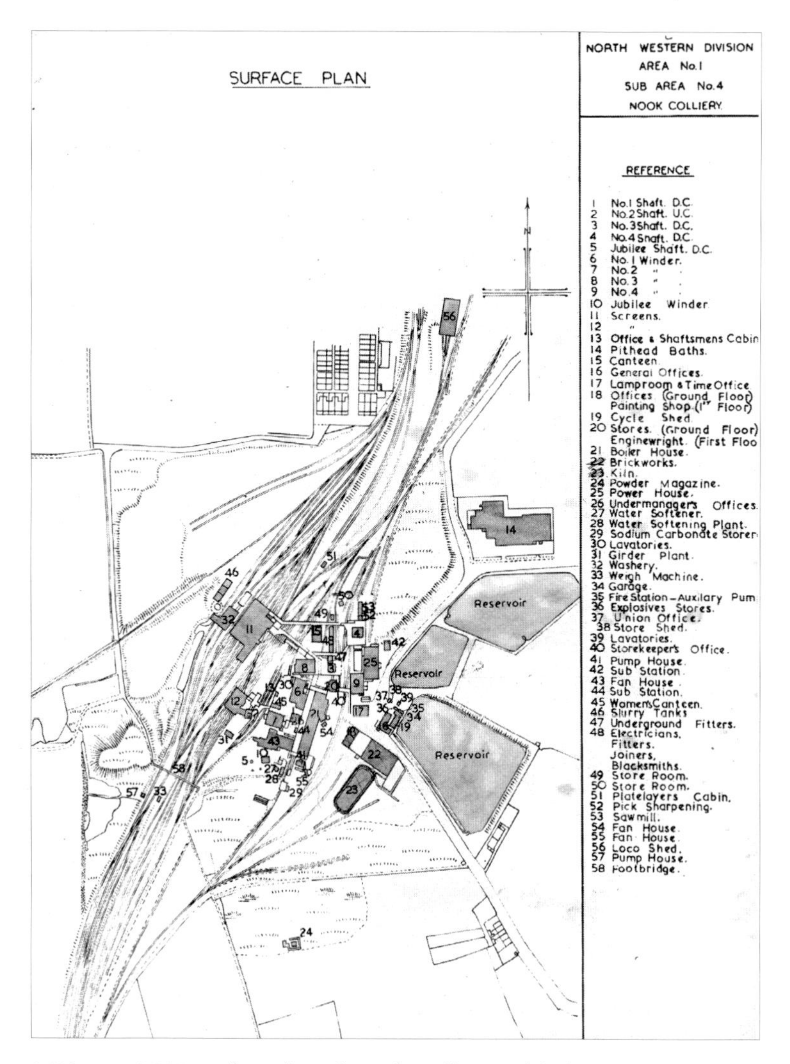

Mid-1950 NCB surface plan of Nook Colliery Tyldesley:
1. No. 1 downcast shaft, 2. No. 2 upcast shaft, 3. No. 3 downcast shaft, 4. No. 4 downcast shaft, 5. Jubilee downcast shaft, 6. No. 1 winder, 7. No. 2 winder, 8. No. 3 winder, 9. No. 4 winder, 10. Jubilee winder, 11. Screen, 12. Screens, 13. Office and shaftsmen's cabin, 14 Pithead baths, 15. Canteen, 16. General offices, 17. Lamproom and time office, 18. Offices and paint shop, 19. Cycle shed, 20. Stores, 21. Boiler house, 22. Brickworks, 23. Kiln, 24. Powder magazine, 25. Power house, 26. Undermanager's house, 27. Water softener, 28. Water softening plant, 29. Sodium carbonate store, 30. Lavatories, 31. Girder plant, 32. Washery, 33. Weigh machine, 34. Garage, 35. Fire station and auxiliary pump, 36. Explosives store, 37. Union office, 38. Store shed, 39. Lavatories, 40. Storekeepers office, 41. Pump house, 42. Sub station, 43. Fan house, 44. Sub station, 45. Women's canteen, 46. Slurry tanks, 47. Underground fitters, 48. Electricians, fitters, joiners, 49. Store room, 50. Store room, 51. Platelayers cabin, 52. Pick sharpening, 53. Sawmill, 54. Fan house, 55. Fan house, 56. Loco shed, 57. Pump house. 58. Footbridge

Outwood

Outwood Colliery, Radcliffe, was sunk around the 1830s as Clough Side Colliery. Outwood Colliery Co. was formed in 1909. One of the new members of Manchester Collieries in March 1929, Outwood was ideally situated, a quarter of a mile down the line from Radcliffe Power Station, but ceased production by February 1931.

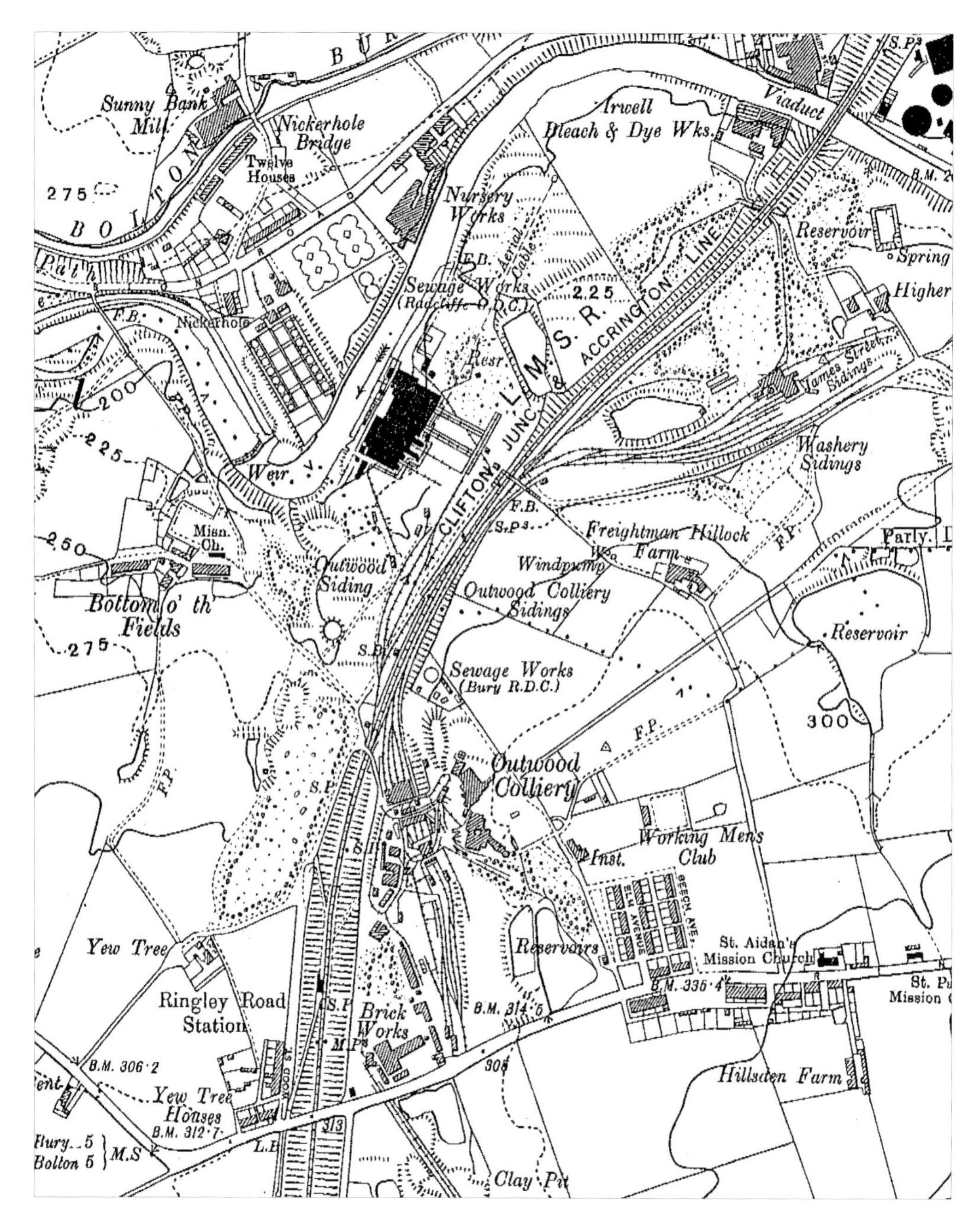

Outwood Colliery (centre) alongside the Clifton Junction & Accrington LMS line in 1931. A few hundred yards (solid black square) to the north is Radcliffe Power Station, closed 1959. A coal washery was operated on site at Outwood by the NCB until 1955.

Sandhole

A poor quality but important aerial view to the north west of Sandhole (Bridgewater Colliery) *c.* 1932 showing the pit alongside the Bridgewater Collieries mineral line, Manchester Rd top right. Note the screens with twin slurry cones at the bottom. The first two shafts were sunk from around 1865, with two more sunk in the 1870s.

Sandhole Colliery from the south-east on 26 September 1964, demolition in progress. Manchester Collieries invested a large amount in the colliery development, the concrete encased headgear of 1935 (left) one of the first in Britain. The colliery closed in September 1962, 1,043 men losing their jobs. (Philip Hindley)

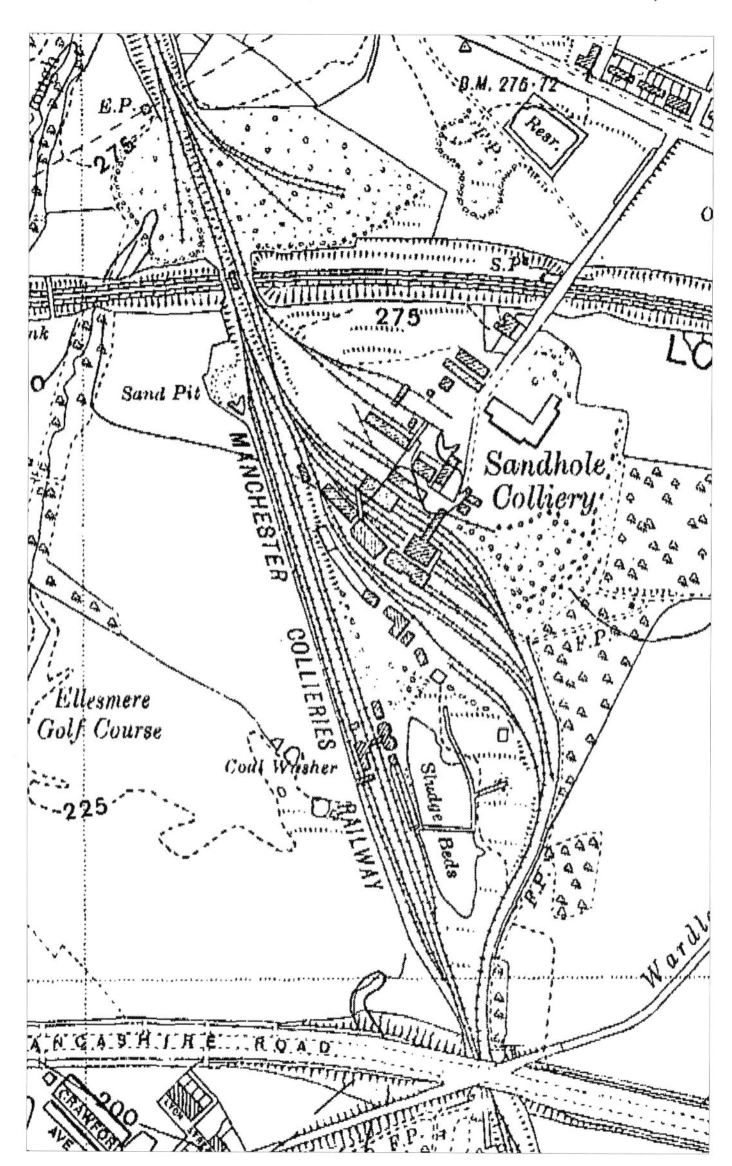

1938 OS plan of the Sandhole Colliery site. The East Lancs Rd bridge can be seen at the bottom where the line headed to Roe Green coal wharf and eventually the Worsley canal coal tip. The LMS Walkden High Level–Manchester Victoria line cuts through the site north of the colliery.

St Georges

St Georges Colliery, Tyldesley, in the distance, 1940. Civil Defence Volunteers exercise in progress. Sunk by Astley & Tyldesley Coal Co. around 1866. Linked below ground to Chanters Colliery, Atherton, by Manchester Collieries. Coal winding ceased in 1941. A miners' training centre and ventilation site until 1964.

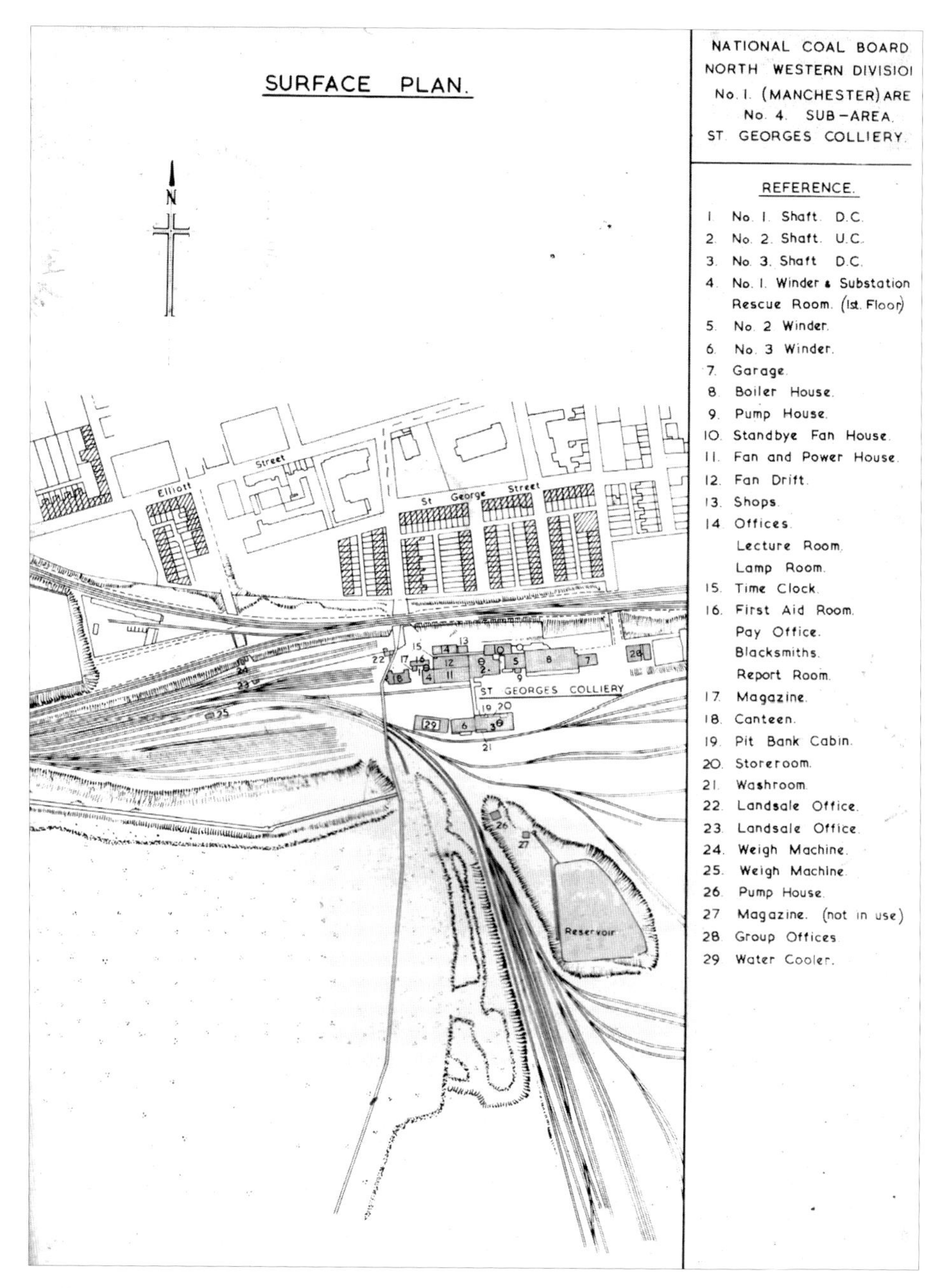

Mid-1950 NCB surface plan of St Georges Colliery:
1. No. 1 downcast shaft, 2. No. 2 upcast shaft, 3. No. 3 downcast shaft, 4. No. 1 winder and sub station, rescue room, 5. No. 2 winder, 6. No. 3 winder, 7. Garage, 8. Boiler house, 9. Pump house, 10. Standby fan house, 11. Fan and power house, 12. Fan drift, 13. Shops, 14. Offices, lecture room, lamp room, 15. Time clock, 16. First aid room, pay office, blacksmiths, report room, 17. Magazine, 18. Canteen, 19. Pit bank cabin, 20. Storeroom, 21. Washroom, 22. Landsale office, 23. Landsale office, 24. Weigh machine, 25. Weigh machine, 26. Pump house, 27. Group offices, 28. Water cooler

Wharton Hall

Wharton Hall Colliery, east of Little Hulton, during construction of the screening shed *c.* 1885. Sunk around 1873 by the Wharton Hall Colliery Co. Bridgewater Collieries purchased the pit in 1879–80, expanding the 'take' of their Brackley Colliery a few hundred yards to the north. A pumping station from 1927 to 1964.

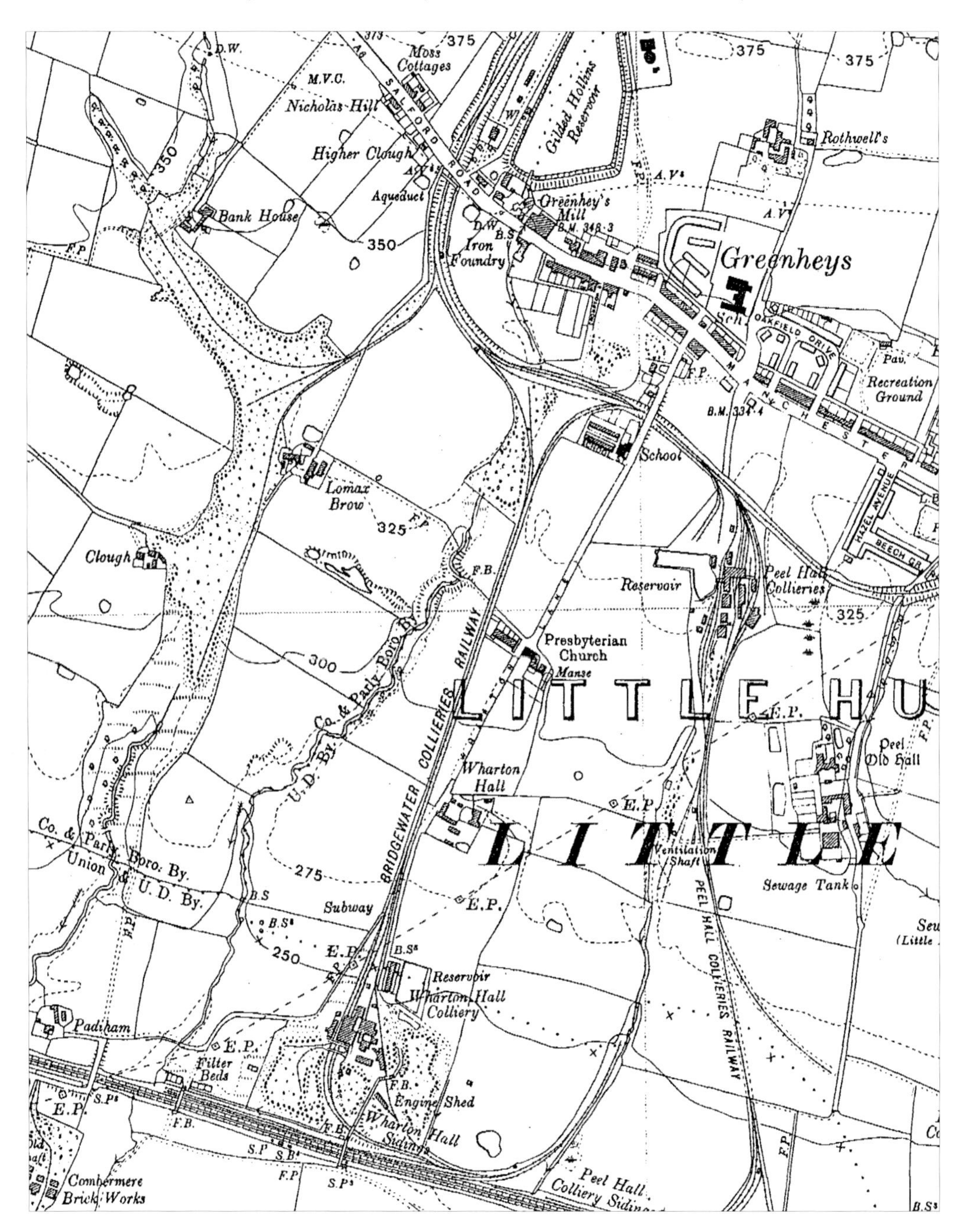

Wharton Hall Colliery is at the lower edge of this 1938 OS plan, adjacent to the LNWR where it had sidings. Bridgewater Collieries laid a line to the north, their Brackley Colliery just off the map at the top. Peel Hall Colliery is middle right. The future enormous Cutacre dirt tip is in its early stages top left.

Wheatsheaf

The two pit headgears of Wheatsheaf Colliery dominate the skyline in Pendlebury on Friday 23 May 1953. Sunk *c.* 1842 by Andrew Knowles & Sons. Coal production at Wheatsheaf ceased on 16 June 1961, 505 men losing their jobs.

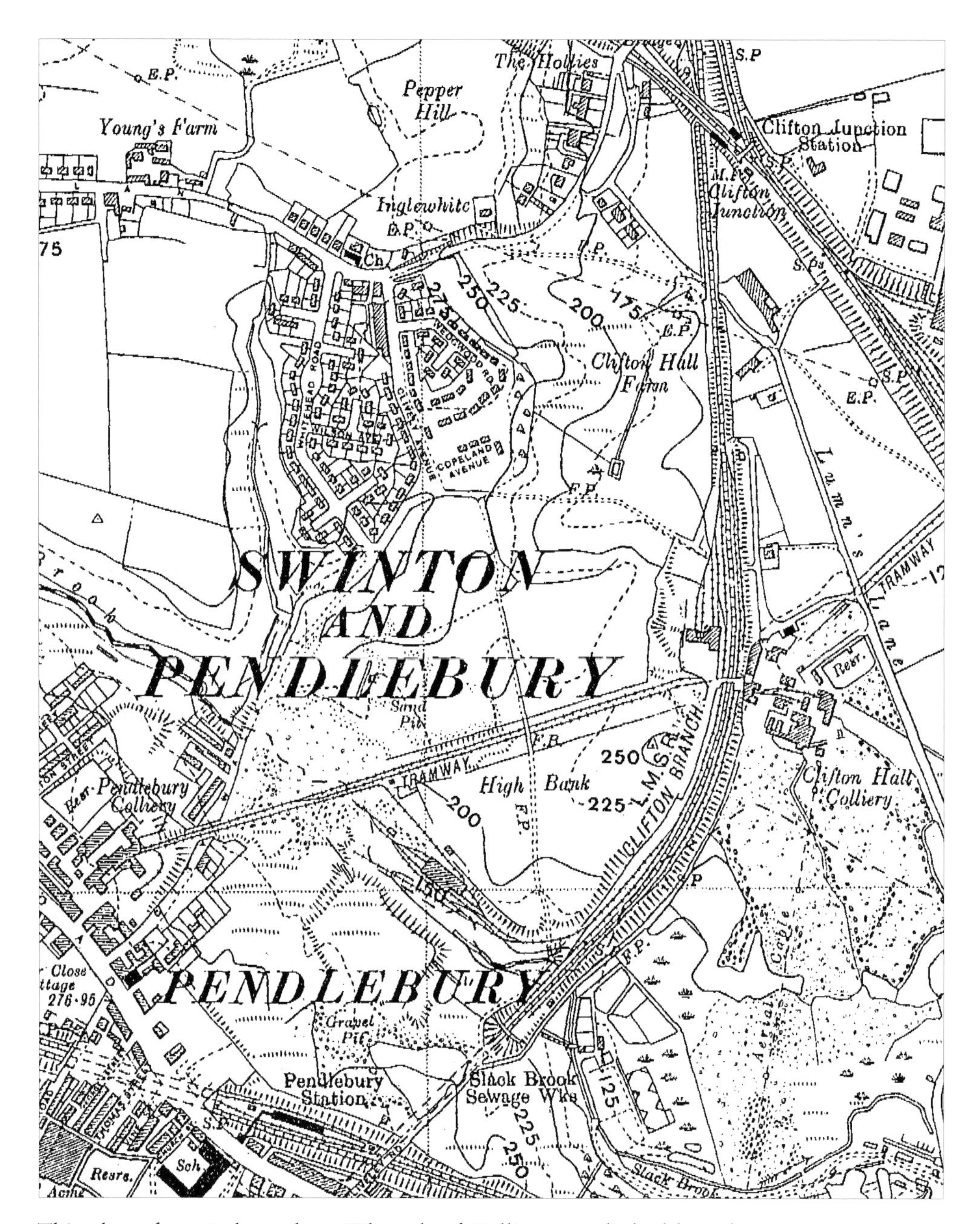

This plan of 1938 shows how Wheatsheaf Colliery was linked by tub tramway to screens adjacent to the LMS Clifton Branch at the former Clifton Hall Colliery site (closed 9 November 1929). Dirt fed the tips at Clifton via aerial ropeways. A very busy coal landsale yard operated at Wheatsheaf.

Appendices

Appendix 1

A list of former Lancashire Coalfield locomotives preserved or undergoing restoration, not exhaustive, April 2013.

Bellerophon (Haydock Foundry 0-6-0WT 1874/C) is in running order at the Keighley & Worth Valley Railway

Gwyneth RSH/7135 no longer exists as a loco, although its boiler and motion work was used for the National Railway Museum broad gauge replica of GWR broad gauge loco *Iron Duke c.* 1984.

Harry (HC/1776) is undergoing restoration (2012–) at the Horwich workshops of Bryn Engineering (John Marrow, ex-Kirkless Workshops, Walkden Yard, and Quaker House Colliery, Wigan, and his son) for use on the Embsay Railway as HE/3776 No. 7. Web site at the time of writing was http://respite3696.tripod.com/Hunslet2414/id11.html

Lindsay (WC&ICo./1887). On loan to the former Steamtown Railway Museum at Carnforth, now West Coast Railways, from 10 March 1981. It can be seen today (2013) by permission.

Princess (NSR 2/1923 LMS, LMS No. 2271, built around the frame of *Sir Robert*). Non-running static exhibit at the National Railway Museum, Shildon.

Repulse (HE/3698) is in superb running order at the Lakeside & Haverthwaite Railway.

Respite (HE/3696) is currently (the remains of) at the Ribble Steam Railway. Riversway Docklands, Preston, Lancashire. Long-term plans are for a rebuild (2013).

Warrior (HE/3823) is at the Dean Forest Railway under restoration.

Whiston (HE/3694) is at the Foxfield Railway in Staffordshire and in steam earlier this year (2013).

'1861' (0-4-0WT Hawthorns, Leith 244/1861) formerly *Ellesmere*, ex-Atherton Collieries is preserved as a static exhibit at The National Museum of Scotland, Edinburgh.

Appendix 2

Extract from an agreement between Manchester Collieries Ltd and the Central Wagon Co. Ltd for the sale of railway wagons dated 3/4/1945 (courtesy of and transcribed by Philip Hindley)

"Whereas Manchester Collieries Ltd. own 8,398 wagons and have hired 1,150 wagons under hire purchase agreements and have agreed to sell the wagons and their rights under the hire purchase agreements to Central Wagon Co. Ltd for the sum of £585,110 of which £526,387 shall be payable in cash and the balance of £58,723 shall represent the outstanding hire purchase liabilities to be taken over by Central Wagon Co. Ltd."

Manchester Collieries Ltd. to pay compensation if:

(1) The wagons or any of them are compulsorily acquired within 10 years of 31/3/1945 under any Act of Parliament or statutory rule or order in council.

(2) Any wagons are condemned or excluded from main line traffic during a period of 10 years from 31/3/1945

Manchester Collieries Ltd. to hire the wagons from Central Wagon Co.Ltd. from 1/4/1945 and will (among other things) keep wagons in good repair in running order and do all necessary greasing and oiling to axles and axle trees.

As soon as practicable after the wagons cease to be requisitioned Manchester Collieries Ltd. shall cease to repair main line wagons, thereafter all the wagons to be repaired by Central Wagon Co.Ltd. at Manchester Collieries Ltd. expense.

1st Schedule – Hire Purchase Agreements (all 10 years)

Wagon Finance Corporation.Ltd.	200	31/5/1936 (date commenced)
Wagon Finance Corporation.Ltd.	50	4/6/1935
Wagon Finance Corporation.Ltd.	50	1/12/1936
Wagon Finance Corporation.Ltd.	175	1/1/1937
Wagon Finance Corporation.Ltd.	150	1/9/1939
Lancashire & Yorkshire Wagon Co.Ltd.	75	12/11/1936
Lancashire & Yorkshire Wagon Co.Ltd.	150	15/9/1939
Chas.Roberts & Co.Ltd.	100	1/12/1936
Chas.Roberts & Co.Ltd.	200	1/10/1939

2nd Schedule – Valuations

Capacity	Year Built	Cost Price	Central Wagon Valuation
8T	All	£30	£20
10T	All	£40	£35
12T	Before 1920	£50	£50
12T	1923/4	£100	£70
12T	1925-7	£110	£75
12T	1928	£110	£80
12T	1934-7	£140	£100
12T	1939	£150	£130

3rd Schedule – Rate of hire per week

8T	All	1/6d
10T	All	2/0d
12T	Before 1920	3/0d
12T	1920-30	3/6d
12T	After 1930	4/3d

Agreement between Manchester Collieries Ltd and The Central Wagon Co.Ltd and The Central Wagon Hiring Co.Ltd dated 1/4/1946

Included:

(1) Number of new wagons subject to 3/4/1945 Agreement is 9530

(2) 9530 wagons sold by Central Wagon Co.Ltd. to Central Wagon Hiring Co.Ltd.

(3) All obligations of Manchester Collieries Ltd. under the earlier agreement "should be and become and should endure" (*quote*) for the benefit of the Central Wagon Hiring Co.Ltd and its assigns in lieu of the Central Wagon Co.Ltd.

Agreement between Manchester Collieries Ltd and The Central Wagon Hiring Co.Ltd dated 24/6/1946

Covered the hire of 9515 wagons for 9 years from 1/4/1946 and included the same figures for valuations and rates of hire as the agreement of 3/4/1945

Bibliography

Includes a list of publications containing information specifically on Walkden Yard and locomotives (courtesy Philip Hindley)

Lancashire Archives, Preston
NCB BW Bridgewater Collieries records

NCB Bw 11/1-2 Valuation of Bridgewater Collieries, etc, for estate duty on the Third Earl of Ellesmere (deceased) as on 13th July 1914

NCB Bw 11/5 Bridgewater Collieries in Reconstruction as at 31st December 1920 (valuation done as additions or deductions to valuation of 13/7/1914)

NCB Bw 19/1 General correspondence, papers and plans of John and James Ridyard, colliery agents, Walkden Moor *1807-1867*

NCB Bw 20/15 Bridgewater Collieries Ltd. Heavy Expenditure Book 1921-28

NCB Bw 20/16 Property ledger. Valuations of structures, plant etc 1921-1930

The Wigan Coalfield, Alan Davies, Tempus, 1999-2009.
The Pit Brow Women of the Wigan Coalfield, Alan Davies, Tempus, 2006.
The Atherton Collieries, Alan Davies, Amberley, 2009
The Pretoria Pit Disaster, A Centenary Account, Alan Davies, Amberley, 2010
Coalmining in Lancashire & Cheshire, Alan Davies, Amberley, 2010
Walkden Yard, Alan Davies, Amberley, 2013
Lancashire Record Office, NCB / Bridgewater Collieries records.
Nef, J. U., *The Rise of the British Coal Industry*. 1932, Routledge, 1952.
North Western Coalfields Regional Survey Report, Ministry of Fuel & Power, HMSO 1945.

Colliery Year Book & Coal Trades Directories, 1943 to 1963.
Guide to the Coalfields. Colliery Guardian, 1948 to 1986.
The Colliery Manager's Handbook, Pamely, Crosby Lockwood *c.* 1904.
Coalmining, G. Preece, City of Salford Cultural Services Dept, 1981.
Collieries in the Manchester Coalfields, Geoff Hayes, Landmark 2004.
Manchester Collieries Ltd newsletter; 1944, 1945, 1946.
Transactions of the Institute of Mining Engineers.
The Colliery Guardian.
The Industrial Railways of the Wigan Coalfield series, parts 1 and 2, The Manchester Coalfield, Bolton & Bury, Wigan. Townley, Appleton, Peden, Smith, Runpast 1994, 1995.
Mines and Miners of South Lancashire 1870-1950, Donald Anderson and J. Lane, pub Donald Anderson 1980.
Coal mining in the Eighteenth and Nineteenth Centuries, B. Lewis, Longmans, London. 1971.
The Miners, In Crisis and War from 1930 onwards. R. Arnot, Allen & Unwin Ltd
Coal – Technology for Britain's Future. Sir Derek Ezra & Others, Macmillan, London, 1976
Coal Mining. Dr A. R. Griffin, pub Longman. 1971.
The Problem of the Coal Mines, Redmayne, R. A. S., 1945.
The History of the British Coal Mining Industry. Vol.2 1700–1830. M. W. Flinn, Oxford U.P, 1984
The History of the British Coal Mining Industry. Vol.3 1830–1913. R. Church, Oxford U.P, 1986
The History of the British Coal Mining Industry. Vol.4 1913–1946. Prof. B. Supple, Oxford U.P, 1987
The History of the British Coal Mining Industry. Vol.5. 1946-1982. W. Ashworth, W. Oxford U. P., 1986
NCB Coal Magazine 1947-1960
Coal - A Pictorial History of the British Coal Industry, Donald Anderson. David & Charles, Newton Abbot 1982
Industrial Railway Record
Porta, L. D. Steam Locomotive Power: Advances Made During the Last Thirty Years. The Future. 1990, 1991.
Railways Illustrated. No. 61.
Railway Magazine. August 1963.
Railway Magazine. January 1990.
Railway Magazine. February 1992.
Townsley, D.H.: *The Hunslet Engine Works.* 1998.
Collieries & Their Railways in the Manchester Coalfields – 2nd Edition. Geoffrey Hayes, Landmark Publishing 2004

Industrial Locomotives of Lancashire; Part A, The National Coal Board including Opencast Disposal Points and British Coal by V. J. Bradley & P. G. Hindley Industrial Railway Society 2000
Walkden Yard 1932-1939 by George L. Booth, Walkden Library Local History Group, Publication No.1

Walkden Yard Yarns No.1, George L. Booth
Walkden Yard Yarns No.2, George L. Booth, (both *c.* 1980s)
The Trustees Railways and After, George L. Booth, 1994
Working On't Railroad 1939-1950, E. A. Brooks, 1991

Industrial Railway Society RECORD Magazine:
No.44 *Joseph* and *Bridgewater* article with photos by G. Hayes
No.172 Destination Walkden Yard by Cyril Golding
No.172 Walkden: The Last Two Years by Steve Leyland
No.196 Modifications to NCB Steam Locomotives by P. G. Hindley, D. Holroyde and S. Oakden (includes details of modifications carried out at Walkden to reduce smoke)

Industrial Locomotive Society THE INDUSTRIAL LOCOMOTIVE Magazine:

No.13 Recollection of the Bridgewater Section of Manchester Collieries by C. A. Appleton
No.41 Austerity Tanks at Collieries in Lancashire by C. A. Appleton
No.42 Austerity Tanks at Collieries in Lancashire – Part 2 by C. A. Appleton
No.43 Austerity Tanks at Collieries in Lancashire – Part 3 by C. A. Appleton
No.45 Austerity Tanks at Collieries in Lancashire by C. A. Appleton (additional notes)
No.47 Austerity Tanks at Collieries in Lancashire by C. A. Appleton (additional notes)
No.60 *Katharine* article with photo by C. A. Appleton
No.62 From Walkden Yard Records Part One by C. A. Appleton
No.63 From Walkden Yard Records Part Two by C. A. Appleton
No.66 Accident at Wharton Hall by C. A. Appleton
No.67 *Katharine* works photo
No.67 *William* of Walkden – Ex-GWR 244 article with photo by C. A. Appleton
No.72 Another Adventure at Wharton Hall by C. A. Appleton
No.77 The Wanderings of *Robin Hood* by C. A. Appleton
No.79 The Wanderings of *Carbon* by C. A. Appleton

The Locomotive Magazine 15 October 1924
'Locomotive for Bridgewater Collieries' – details of loco *Bridgewater* with photo

The Railway Magazine March 1963
'The "Knotty" Steams On' – article on ex-North Stafford 0-6-2T locos

ALSO AVAILABLE FROM AMBERLEY PUBLISHING

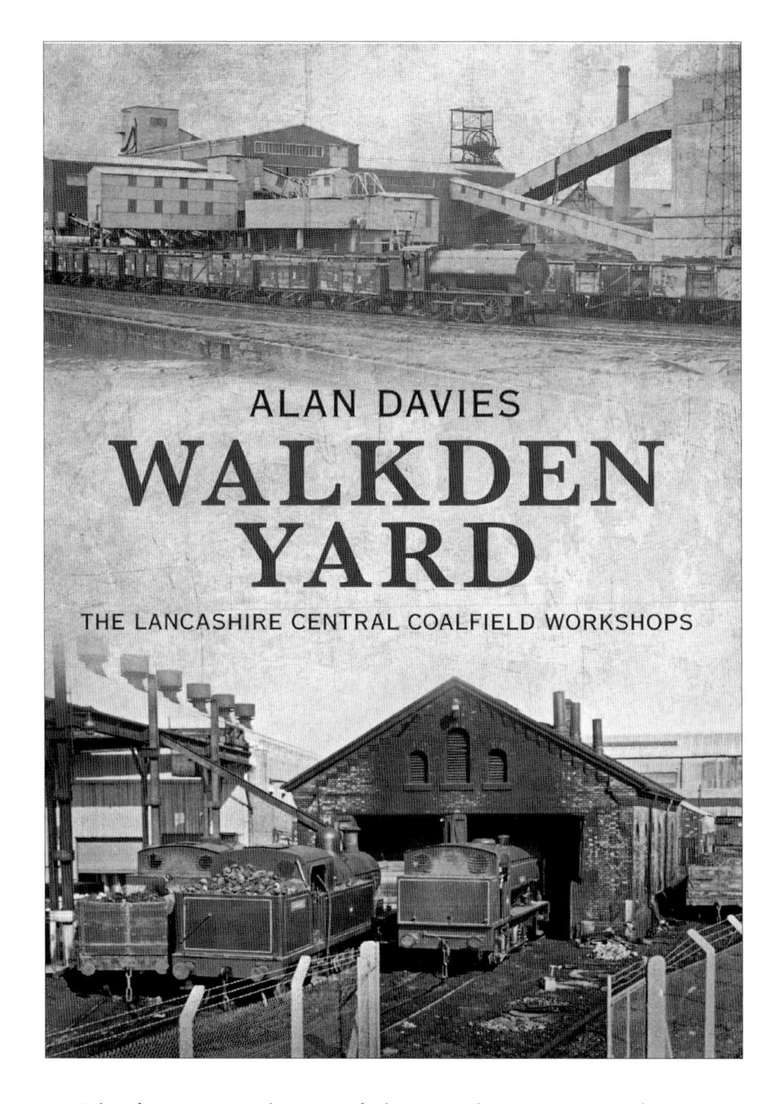

This fascinating selection of photographs gives an insight into the history of the Walkden Yard and the Lancashire Coalfield Workshops.

978-1-84868-925-1
128 pages, illustrated throughout

Available from all good bookshops or order direct from our website www.amberleybooks.com

ALSO AVAILABLE FROM AMBERLEY PUBLISHING

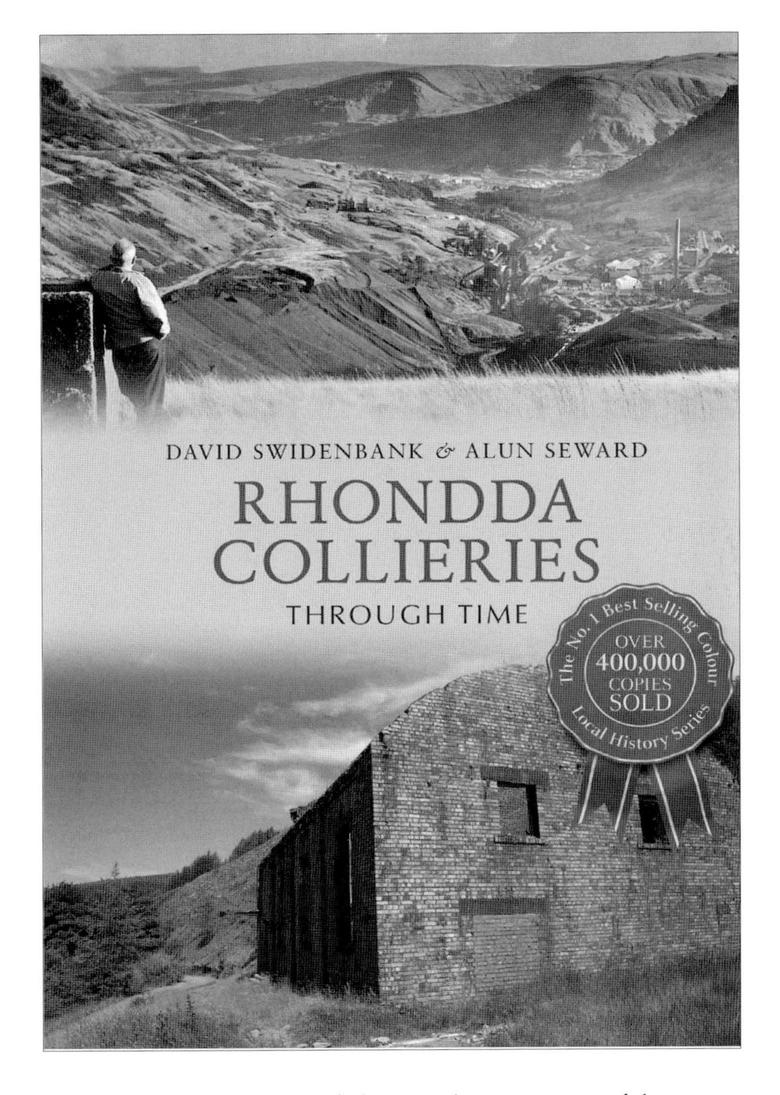

This fascinating selection of photographs traces some of the many ways in which Rhondda Collieries have changed and developed over the last century.

978-1-4456-1064-1
96 pages, illustrated throughout

Available from all good bookshops or order direct from our website www.amberleybooks.com

ALSO AVAILABLE FROM AMBERLEY PUBLISHING

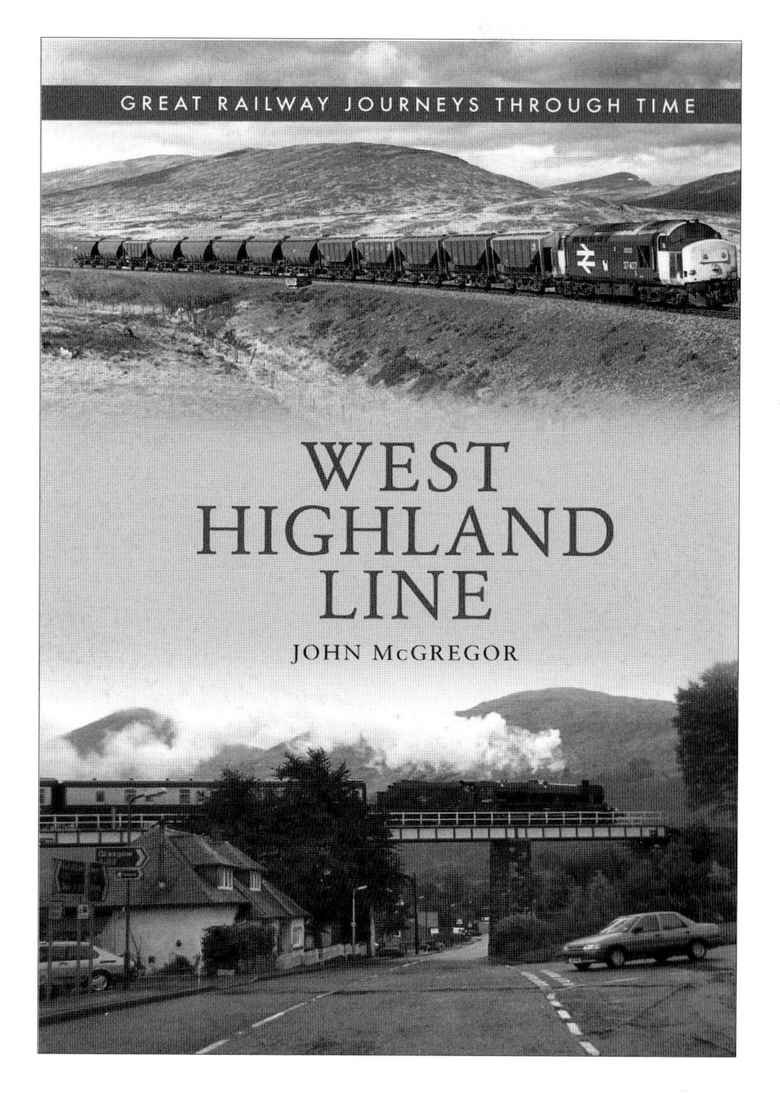

This fascinating selection of photographs gives an insight into the history and landscapes of the West Highland Line.

978-1-4456-1336-9
96 pages, 180 illustrations

Available from all good bookshops or order direct from our website www.amberleybooks.com